Project Management for Research
A guide for engineering and science

Adedeji B. Badiru
School of Industrial Engineering
University of Oklahoma
Oklahoma
USA

Published by Chapman & Hall, 2–6 Boundary Row, London SE1 8HN, UK

Chapman & Hall, 2–6 Boundary Row, London SE1 8HN, UK

Blackie Academic & Professional, Wester Cleddens Road, Bishopbriggs, Glasgow G64 2NZ, UK

Chapman & Hall GmbH, Pappelallee 3, 69469 Weinheim, Germany

Chapman & Hall USA, 115 Fifth Avenue, New York, NY 10003, USA

Chapman & Hall Japan, ITP-Japan, Kyowa Building, 3F, 2-2-1 Hirakawacho, Chiyoda-ku, Tokyo 102, Japan

Chapman & Hall Australia, 102 Dodds Street, South Melbourne, Victoria 3205, Australia

Chapman & Hall India, R. Seshadri, 32 Second Main Road, CIT East, Madras 600 035, India

First edition 1996

© 1996 Adedeji B. Badiru

Typeset in 11/13 Times by Best-set Typesetter Ltd., Hong Kong

Printed in Great Britain by T. J. Press, Padstow, Cornwall

ISBN 0 412 58890 0

A catalogue record for this book is available from the British Library

Library of Congress Catalog Card Number: 95-74653

Contents

To Jane Smith, Lisa Robinett and Jean Shingledecker, who continue to provide a conducive and supportive office environment.

Preface

Graduate research is a complicated process which many engineering and science students aspire to undertake. The complexity of the process can lead to failures for even the most brilliant students. Success with graduate level research requires not only a high level of intellectual ability, but also a high level of program management skills.

After many years of supervising several graduate students, I have found that most of them have the same basic problems of planning and implementing their research programs. Even the advanced graduate students need the same 'mentoring and management' guidance that has little to do with actual classroom performance. It is my conjecture that graduate students could make a better job of their research programs if a self-paced guide were available to them.

The guide provided in this book covers topics ranging from how to select an appropriate research problem to how to schedule and execute research tasks. The book takes a project management approach to planning and implementing graduate research in engineering, science and manufacturing disciplines. It is a self-paced guide that will help graduate students and advisors answer most of the basic questions about 'how to do this and how to do that'. There is a need for such a guide book. The book will alleviate frustration on the part of the student and the research advisor. If a graduate student is organized and prepared to handle the basic research management functions, he or she and the advisor will have more time for actual intellectual mentoring and knowledge transfer.

Acknowledgments

The inputs and ideas of the students over the years were instrumental in the composition of several sections of the book. I thank my present and former graduate students for creating the opportunities that made it possible for me to write this book: Joseph Chetupuzha, Rajesh Nath, Karode Amol, Gaugarin Oliver, Ravindra Sunku, Krishna Muppavarapu, Sundaram Deepak, Arif Alaa, Radha Maganty, Massy Nakada, Neetin Datar, Revathi Advaithi, Hemanth Jayaraman, Linda Fox, Herschel Baxi, Pamela McCauley-Bell, Steve Rogers, John Peters, Cathy Modaro, and Dave Sieger. I specially thank Volker Baukmann, my German exchange student, Stanislav Vasilev, Tong Vasilev, Milan Milatovic and Ibrahim Al-Harkan for their intellectual inputs about graduate education systems around the world. I also thank my undergraduate research assistants, Marcus Stocco, Jonathan Hope, Lloyd Martin and Abi Badiru.

Special thanks go to our graduate secretary, Lisa Robinett, whose insightful comments, administrative help and interest in the success of our graduate students provided motivation for me to write this book. As always, I thank my family for bearing with me while I embarked on yet another lengthy book writing project.

Graduate education process | 1

'Every noble work is at first impossible.' Benjamin Franklin

Graduate education can be very complex and demanding. Surviving in this highly stressful environment takes a mixture of intelligence, self-discipline and organization. This book presents guidelines for graduate education and research. There is a technical requirement to coping with graduate research and there is also a managerial requirement. While much has been said and written about the technical requirements, very few guidelines are available for the management of the process.

This book presents a managerial approach to graduate education and research for engineering and science. The proven techniques of formal project management are recommended as effective approaches to achieving the objectives of graduate research. The book is deliberately written in an informal list-oriented style to facilitate ease of reference. It is intended to assist both students and faculty members in the pursuit of the activities necessary for satisfactory completion of graduate research in science and engineering. The pursuit of graduate education is an individual matter. It is the responsibility of the student to take the appropriate action at the appropriate time.

1.1 PECULIARITIES OF GRADUATE EDUCATION

While attempting to make the most out of student–advisor inter-actions, I have found that many graduate students lack the basic elements that facilitate good graduate study. Unfortunately there are very few formal avenues for teaching graduate students how to manage their academic programs properly. I have always wished that I could refer my students to some publication where they could find the necessary guidelines. There are all types of courses available for topics such as money management, time management and personal management, but none in academic management. The guidelines presented in this book can help fill that void.

Graduate education requires students with professional maturity who are motivated and committed to learning. They must play an active role in their education and understand the value of learning as a life-long professional goal. They must also have a clear perception of their career and educational objectives.

Unfortunately, many graduate students lack the skills necessary to plan and implement their research programs independently. Graduate advisors often have difficulty finding a good reference book to suggest to graduate students concerning basic procedures for implementing their research plans.

> You must approach your graduate program with dedication and utmost commitment.

There are several peculiar characteristics of graduate education that are not found in other academic levels. The research component of graduate education makes it particularly petrifying. Some of the unique aspects of graduate education that students must become aware of are:

- library etiquette;
- personal responsibility;
- effective consultation with research advisors;
- preparations for graduate advisory conferences;
- beating academic deadlines;
- allowing enough lead time for activities;
- planning, scheduling, and controlling research activities;
- managing independent research studies;
- preparation of good research reports;
- familiarization with graduate research policies and procedures;
- interaction with other graduate students;
- formal and informal interactions with research advisors;
- involvement with student chapters of professional organizations.

Most students entering graduate schools are assumed to already have a good level of technical preparation. Where they need help is with being organized and managing themselves well. Essential characteristics for success with graduate education include:

- intelligence
- ingenuity
- creativity
- conducive environment

Organize, organize, and organize again!

Some of the first and foremost requirements for success include:

- Organize, organize, organize.
- Know where everything is kept.
- Standardize your own filing system and techniques so that you can find needed information.

After directing several graduate students over ten years, I have found that they all have the same basic questions. Even the advanced students ask the same basic questions. A general guide book is essential to assuring consistency in graduate research supervision. It is suspected that the same advisor will give ten different answers to the same question asked by the same student at ten different times. Conversely, ten different advisors will give ten different answers to the same question asked by the same student.

Knowledge and intelligence alone are not enough to succeed with graduate education. Graduate students must have the skills necessary to apply knowledge and the attitude required to be a responsible and ethical professional. Skills such as those required for effective learning, problem solving and communication cannot be learned in the traditional lecture format. Graduate students must exhibit the motivation, self-discipline, and dedication for doing more outside class to enhance their graduate education.

Why graduate education? For several reasons:

- because the world is not perfect;
- because new technologies evolve from advanced studies;
- because more years of study can change a perspective;
- because education makes people better;
- because the world cannot get stale if we keep learning;
- because it facilitates employment;
- because it provides new ways of getting things done;
- because it brings more social pride.

1.2 PREPARING YOURSELF FOR GRADUATE EDUCATION

- Proficiency in the following computer skills will be used throughout the graduate research:
  - (a) word processing skills;
  - (b) spreadsheet skills;
  - (c) database skills;
  - (d) computer programming languages;
  - (e) equation solving skills;
  - (f) graphic representation skills for data analysis and results;
  - (g) presentation software skills.
- Proficiency with research tools and strategies includes:
  - (a) problem definition;
  - (b) search and use of available literature;
  - (c) use of the library and other information sources;
  - (d) supporting analysis such as data analysis, statistical analysis, economic analysis, and research safety.
- Development of the following problem solving skills will be useful in open-ended problems that may be encountered during the research:
  - (a) developing a vision for a solution;
  - (b) extensive reading habit;
  - (c) rigorous course work;
  - (d) participation in class discussions.
- The development of written communication skills includes the following, which are particularly related to technical reports for disseminating research results:
  - (a) developing draft versions;
  - (b) editing and proofreading;
  - (c) using editors and proofreaders;
  - (d) developing final drafts.
- The development of interpersonal skills will be required for working in large research teams.
- The development of a sense of responsibility and ethics in science and engineering is necessary for a professional career.
- Familiarity with the following communication tools will be useful:
  - (a) electronic mail;
  - (b) fascimile (FAX);
  - (c) multimedia;
  - (d) telecommuting;

    (e) tele-researching;

    (f) remote interface with research facilities and research advisors.

- You will need to assess your own motivation, drive, objectives, and academic preparation.

The student must demonstrate ability to do original and creative research of a caliber that will advance the state of knowledge in a science or engineering field. A combination of a problem-based and case-based method of graduate education is the primary approach to successful graduate research. This fosters a self-directed learner who is able to integrate, apply and communicate changing and expanding information.

1.3 PROJECTING AND PRESERVING GOOD PERSONAL IMAGE

Good personal image is essential to pursuing advanced degrees in science and engineering. Self-motivation and commitment will enhance your personal image. Your needs as an individual should be taken into account in your educational preparations. The human hierarchy of needs suggests the following categories as possible drivers for graduate education:

1. **Physiological needs**: the needs for the basic things of life, such as food, water, housing and clothing. This is the level where access to money is most critical.
2. **Safety needs**: the needs for security, stability and freedom from threat of physical harm. The fear of adverse environmental impact may inhibit project efforts.
3. **Social needs**: the needs for social approval, friends, love, affection and association. For example, public service projects may bring about better economic outlook that may enable individuals to be in a better position to meet their social needs.
4. **Esteem needs**: the needs for accomplishment, respect, recognition, attention and appreciation. These needs are important not only at the individual level, but also at the national level.
5. **Self-actualization needs**: these are the needs for self-fulfillment and self-improvement. They also involve the availability of opportunity to grow professionally. Work improvement projects

may lead to self-actualization opportunities for individuals to assert themselves socially and economically. Job achievement and professional recognition are two of the most important factors that lead to employee satisfaction and better motivation.

Hierarchical motivation implies that the particular motivation technique utilized for a given person should depend on where the person stands in the hierarchy of needs. For example, the needs for esteem take precedence over the physiological needs when the latter are relatively well satisfied. Money, for example, cannot be expected to be a very successful motivational factor for an individual who is already on the fourth level of the hierarchy of needs.

1.3.1 Educational motivators

Motivation can involve the characteristics of the education process itself. In the theory of motivation, there are two motivational factors classified as the hygiene factors and motivators. Hygiene factors are necessary but not sufficient conditions for a contented individual. Negative aspects may lead to disgruntlement, whereas positive aspects do not necessarily enhance personal satisfaction. Examples include:

- **Administrative policies**: bad policies can lead to discontent while good policies are viewed as routine and make no specific contribution to improving satisfaction.
- **Supervision**: a bad supervisor can make a person unhappy and less productive while a good supervisor cannot necessarily improve performance.
- **Working conditions**: bad working or study conditions can impede students, but good conditions do not automatically generate improved productivity.
- **Salary**: low salary can make a person unhappy, disruptive and uncooperative, but a raise will not necessarily induce him or her to perform better. While a raise in salary will not necessarily increase professionalism, a reduction in salary will most certainly have an adverse effect on morale.
- **Personal life**: a miserable personal life can adversely affect performance, but a happy life does not imply better performance.
- **Interpersonal relationships**: good peer, superior, and subordinate relationships are important to keeping a person happy and pro-

ductive, but extraordinarily good relations do not guarantee that he or she will be more productive.

- **Social and professional status**: low status can force a person to perform at his or her 'level' whereas high status does not imply performance at a higher level.
- **Security**: a safe environment may not motivate a person to perform better, but an unsafe condition will certainly impede productivity.

Motivators are motivating agents that should be inherent in the work or study environment. If necessary, work should be redesigned to include inherent motivating factors. Some guidelines for incorporating motivators into jobs are:

- **Achievement**: the job design should give consideration to opportunity for achievement and avenues to set personal goals to excel;
- **Recognition**: the mechanism for recognizing superior performance should be incorporated into the job design. Opportunities for recognizing innovation should be built into the job;
- **Work content**: the work content should be interesting enough to motivate and stimulate the creativity of the individual. The amount of work and the organization of the work should be designed to fit a person's needs;
- **Responsibility**: an individual should have some measure of responsibility for how his job is performed. Personal responsibility leads to accountability which yields better performance;
- **Professional growth**: the work should offer an opportunity for advancement so that the individual can set his own achievement level for professional growth within a study plan.

The above examples may be described as job enrichment approaches with the basic philosophy that work or study can be made more interesting in order to induce an individual to perform better.

1.4 UNIQUE ASPECTS OF SCIENCE AND ENGINEERING EDUCATION

Science and engineering education have unique characteristics that extend to professional practice. Graduates of science and engineering programs are technical people who have special needs.

Some of these needs are often not appreciated by peers, superiors or subordinates. The unique expectations of technical professionals in terms of professional preservation, professional peers, work content and hierarchy of needs are factors that can influence the graduate education process.

'Professional preservation' is the desire of a technical professional to preserve his or her identification with a particular job function. In many situations, preservation is impossible as a result of a lack of staff to fill specific job slots. It is common to find people trained in one technical field holding assignments in an allied field. An incompatible job function can easily lead to insubordination, egotism and rebellious attitudes. While there will sometimes be a need to work outside one's profession in any job environment, every effort should be made to match the surrogate profession as closely as possible to the employee's own. In most technical training programs, the professional is taught how to operate in the following ways:

- by making decisions based on the assumption of certainty of information;
- by developing abstract models to study the problem being addressed;
- by working on tasks or assignments individually;
- by quantifying outcomes;
- by paying attention to exacting details;
- by thinking autonomously;
- by generating creative insights to problems;
- by analyzing systems operatability rather than profitability.

However, in the business environment, not all of the above characteristics are desirable or even possible. For example, many business decisions are made with incomplete data. In many situations it is simply unprofitable to expend the time and efforts to obtain perfect data. Many operating procedures are guided by company policies rather than the creative choices of employees. The training of science and engineering students, particularly at the graduate level, should prepare them to adapt to different professional practice environments.

Technical people are more likely to be at a higher stage in the hierarchy of needs. Most of their basic necessities for a good life will already have been met. Their prevailing needs will tend to be for esteem and self-actualization. As a result, a technical professional will have expectations that cannot usually be quantified in monetary terms. This is in contrast to non-professionals who may look forward to overtime pay or other monetary remunerations. Technical professionals will generally look forward to achieving one or several of the following:

- **Professional growth and advancement**: professional growth is a primary pursuit of most technical people. For example, a computer professional has to be frequently exposed to challenging situations that introduce new technology developments and enable him to keep abreast of his field. Even occasional drifts from the field may lead to the fear of not keeping up and being left behind by peers. The work environment must be reassuring to technical professionals with regard to the opportunities for professional growth in terms of developing new skills and abilities.
- **Technical freedom**: technical freedom, to the extent permissible within an organization, is essential for the full utilization of a technical background. A technical professional will expect to have the liberty of determining how the objective of an assignment can best be accomplished.
- **Respect for personal qualities**: technical people have profound personal feelings despite the mechanical or abstract nature of their job functions. They will expect respect for their personal qualities. In spite of frequently operating in professional isolation, they do engage in interpersonal activities. They want their non-technical views and ideas to be recognized and evaluated based on merit. An appreciation of their personal qualities gives them the sense of belonging and makes them productive members of a professional team.
- **Respect for professional qualification**: a professional qualification in science and engineering usually takes several years to achieve and is not likely to be compromised by any technical professional. Technical professionals cherish the attention their technical background attracts. They expect certain preferential treatment. They like to make meaningful contributions to decisions. They sometimes take approval of their technical approaches for granted.
- **Increased recognition**: increased recognition is expected as a by-product of a professional effort. Technical professionals view their participation in a project as a means of satisfying one of

their higher-level needs. They expect to be praised for the success of their efforts. They look forward to being invited for subsequent technical endeavors. They savor hearing the importance of their contributions being related to their peers.

- **New and rewarding professional relationships**: new and rewarding professional relationships can serve as a bonus for a project professional. Most technical developments result from joint efforts of people that share closely allied interests. A technical professional will expect to meet new people with whom he or she can exchange views, ideas, and information. The training and practice environments for science and engineering students should be conducive to professional interaction.

1.5 ESTABLISHING EDUCATIONAL GOALS AND OBJECTIVES

Avoid goal conflict and inconsistency in your educational plans.

An educational goal consists of a detailed description of the overall pursuit and expectations from an academic pursuit. A goal is the composite effect of a series of objectives. Each objective should be defined in relation to the career goal of the student. Goal analysis helps to determine what courses to take, what major to choose and what research options to explore.

A goal-clarification approach should be used to set educational goals and objectives. This approach focuses on identifying specific goals and objectives that will assure success. Implementing objectives, tracking performance over time, and providing an ongoing assessment of strengths and weaknesses can help students set and enhance their goals. Techniques such as management by objectives and goal structuring can be used during the education-planning process.

Adequate training is essential because it prepares students to do their jobs well by building the right knowledge that permits logical actions and decision making. If the right skills are provided for people, they can develop efficient work habits and positive attitudes that promote cooperation and teamwork for graduate research and professional practice. Some approaches to training are:

- formal education in an academic institution;
- on the job training through hands-on practice;

- continuing education short courses;
- training videos;
- group training seminars.

Some important aspects of setting educational goals and objectives are:

- to appraise existing categories and levels of technical skills;
- to consider the skills that will be needed in the future;
- to assess local and foreign training opportunities;
- to establish vocational training centers;
- to screen and select candidates for required skill training;
- to monitor the progress of the trainees;
- to synchronize the inflow and outflow of trainees with job potentials and national needs;
- to place trained manpower in relevant technical job functions;
- to monitor the performance of the technical professionals;
- to use the feedback of the working technical professionals as inputs for planning future training programs.

1.6 KNOWLEDGE-DRIVEN VERSUS INCOME-DRIVEN EDUCATION

The pursuit of education may be driven by the need for knowledge or the need for income potential. In each case, the education objectives must be compatible with the characteristics of the individual. Many changes are now occurring in academic programs in preparation for the 21st century. Are we ready? Is the graduate education process ready? As the business world prepares to meet the technological challenges of the 21st century, there is a need to focus on the people who will take it there. People will be the most important component of the 'man–machine–material' systems that will compete in the next century. Graduate and undergraduate educators should play a crucial role in preparing the work force for the 21st century through their roles as initiators and facilitators of change. Improvements are needed in undergraduate education to assure solid graduate education.

Undergraduate education is the foundation for professional practice. Undergraduate programs are the basis for entry into graduate schools and other professional fields. To facilitate this transition, curriculum and process improvements are needed in education strategies. Several educators have recognized that the way engineering and science are practised has changed dramatically over the years and an upgrade is needed to match the times. Educators, employers and practitioners advocate better integration of science with the concepts of design and practice throughout the engineering and science curriculum. Such an integration should be a key component of any education reform in preparation for the 21st century.

Hurried attempts to improve education are being made in many areas. We now have terms like 'Total quality management for academia', 'Just-in-time education', 'Outcome based education', and 'Continuous education improvement'. Unfortunately, many of these are mere rhetoric that are not backed by models of practical implementation. Educators and administrators need guidelines to transform improvement rhetoric and slogans into action.

1.7 QUALITY IN ENGINEERING AND SCIENCE EDUCATION

Incorporating quality concepts into education is a goal that should be pursued at national, state, local, and institution levels. Existing models of total quality management (TQM) and continuous process improvement (CPI) can be adopted for curriculum improvement. However, because academia is unique, TQM will need to be redefined to make it compatible with the academic process. The basic concepts of improving product quality are applicable to improving any education process. Careful review of engineering and science curricula will reveal areas for improvement. This will help avoid stale curricula that may not adequately meet current needs of the society.

1.7.1 Marriage of theory and practice

Teaching is the crux of research while research is the crux of teaching. Integration of teaching and research is required for effective professional practice. The need to incorporate some aspect of practice into engineering education has been addressed widely in the literature. Professors must combine research interests with teaching responsibilities. The declining standards in education can be attributed to waste, lax academic standards, and mediocre teaching and scholarship. The following specific symptoms of educational problems have been cited in the literature:

- increasing undergraduate attrition despite falling academic standards at many schools;
- decreasing teaching loads in favor of increasing dedication to research;
- migration of full professors from undergraduate teaching to graduate teaching and research;
- watered down contents of undergraduate courses in the attempt to achieve retention goals;
- decreasing relevance of undergraduate courses to real-world practice. Figure 1.1 presents a design of a course relevance matrix. The multilayer structure of the matrix permits course linkages across curricula in a fashion that permits the assessment of the relevance of a course to real-world practice.

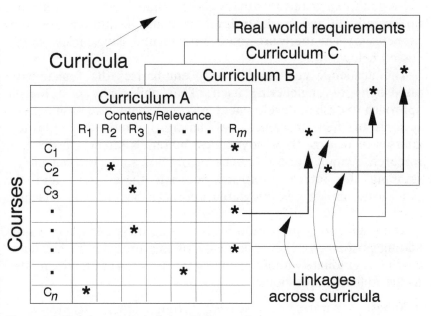

Figure 1.1 Relevance matrix for assessing course content.

1.8 CURRICULUM INTEGRATION

Curriculum integration (an interdisciplinary approach) should be used to address the problems cited above. Curriculum integration should be a priority in reforming education programs. Students must understand the way the world around them works and be capable of becoming responsible contributors to the society. Interdisciplinary education offers a more holistic approach to achieving this goal. Interdisciplinary course and curriculum improvement should link separate but related subjects to provide students with comprehensive skills so that they can adapt to the changing world. One form of interdisciplinary integration involves projects where students from more than one academic department participate in joint industrial projects or on-campus research programs. This facilitates the sharing of views from different angles.

Enhanced engineering education will prepare students to lead efforts to integrate entities in the manufacturing and service organizations of the 21st century. The engineering profession, as a whole, faces an important challenge in educating future engineers for this leadership role. The current engineering curriculum provides good exposure to its many facets. Individual courses at both undergraduate and graduate levels in many institutions are comprehensive. Yet there are some fundamental deficiencies, as discussed below.

The academic curriculum rarely emphasizes the fundamental philosophy of engineering itself. That philosophy is a holistic approach to design, development and implementation of integrated systems of men, machines and materials. Students go through courses in design, thermodynamics, materials science, operations research, manufacturing, human factors and so on without understanding the interrelationships between these areas and the synergistic impact this integrated approach has on man–machine systems.

The engineering profession is also losing its strength as a value-adding profession. The basic cause of this problem is that many engineers graduate without resolving the question of identity related to the following questions:

- What separates engineering from other science-based professions?
- What contribution does the profession make to an organization?

The root of this identity problem lies in the structured and isolated approach of various engineering courses. This results in too narrow specialization. For example, today's graduates tend to associate more with focused professional societies than the general engineering practice.

There is a big difference between the academic and industrial approaches to performance evaluation. This is because the academic community evaluates its members by the number of publications and research grants. In contrast, industry measures performance in terms of real contributions to organizational goals. This has had a detrimental effect on the learning interaction that faculty and students must share in order for students to graduate with professional loyalty, technical competence and the capability to integrate theoretical concepts and industrial practice effectively.

In the attempt to prepare students for graduate level education, the academic curriculum often has a strong mathematical orientation. Though necessary, this develops a structured approach to problem solving among engineering and science students. As a result, students expect all problems to have well-defined inputs, processing modules and outputs. Thus, when faced with the complex, ill-defined and unstructured problems that are common in the real world, many new graduates perform poorly. It is often suggested that the bulk of teaching should be done at the undergraduate level because a large segment of the group go into industry, not to graduate school. Unfortunately, attempts to improve the curriculum are often tilted in favor of research-oriented education, thus depriving the majority of the students of the skills they need to advance in the business world.

Many young graduates mistakenly perceive their expected roles as being part of the management personnel, having little or no direct association with shop floor activities. Such views impede 'hands-on' experience and prevent the identification of root causes of industrial problems. This leads to the development of short-term, unrealistic and/or inadequate solutions. The growing reliance on simulation models that cannot be practically validated in real-world settings is one obvious symptom of this problem.

Some engineering curricula are developed within an isolated shell. Students do not realize the importance of developing solutions that are beneficial to a system rather than to individual components. New graduates often take a long time to become productive in developing solutions that require multidisciplinary science and engineering approaches.

1.9 ETHICS IN EDUCATION

Professional morality and responsibility should be introduced early to engineering students. Lessons on ethics should be incorporated into curriculum improvement approaches. Engineering graduates should be familiar with an engineering code of ethics so that they can uphold and advance the integrity, honor and dignity of their professions by:

- using their knowledge and skill for the enhancement of human welfare;
- being honest, loyal, and impartial in serving the public, their employers and clients;
- striving to increase the competence and prestige of their professions;
- supporting and participating in the activities of professional and technical societies.

1.9.1 Improvement strategy

Some points to consider when developing curriculum improvement approaches are listed below:

- Education should not just be a matter of taking courses, getting grades and moving on. Life-long lessons should be a basic component of every education process. These lessons can only be achieved through a systems view of education. The politics of practice should be explained to students so that they are not shocked and frustrated when they move from the classroom to the boardroom.
- Universities face a variety of real-world multidisciplinary problems that are often similar to industrial operations problems. These problems could be used as test cases for solution approaches. Engineering and science students could form consulting teams and develop effective solutions to such problems.
- Schools should increase their interaction with local industries when such industries are available. This will facilitate more realistic and relevant joint projects for students and industry professionals.
- The versatility of the engineering profession can be enhanced by encouraging students to take more cross-disciplinary courses in disciplines such as computer science, business and human relations.
- Students must keep in mind that a computer is just a tool and not the solution. For example, a word processor is a clerical tool that cannot compose a report by itself without the creative

writing ability of the user. Likewise, a spreadsheet program is an analytical tool that cannot perform accurate calculations without accurate inputs from the user.

- Undergraduate and graduate education should be seen as contiguous components in the overall hierarchy of the education process.

1.9.2 Curriculum assessment

Performance measures and benchmarks are needed to assess the effectiveness of engineering education. The effectiveness of curricula can be measured in terms of the outgoing quality of students. This can be tracked by conducting surveys of employers to determine the relative performance of graduates. The feedback model presented in Figure 1.2 can be used for that purpose.

The variables in the figure are defined as follows:

$I(t)$ = set of inputs (e.g. data, information, technical skill, funding, high school preparation, faculty qualification)

(t) = time reference for performance assessment

$A(t)$ = feedback loop actuator (this facilitates flow of inputs to various segments of the academic system, e.g. functions performed by administrative staff)

$O(t)$ = output of the system in terms of quality of graduates

$G(t)$ = forward transfer process (this coordinates input information and resources to produce output)

$H(t)$ = feedback control process (this monitors the status of improvement and generates feedback information)

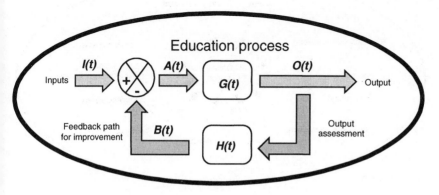

Figure 1.2 Curriculum improvement feedback model.

$B(t)$ = feedback information (this is used to determine control actions needed at the next improvement stage; such information can be used for benchmarking to establish requirements for further improvement)

The primary responsibility of a curriculum improvement team is to ensure proper forward and backward flow of information and knowledge between the academic institution and industry. The percentage of students passing professional registration exams can also be used as a performance measure. The percentage of students progressing to graduate programs and staying to graduate will also be a valuable measure of performance. Entrance questionnaires and exit questionnaires can be used to judge students' perception of the improved curriculum.

1.10 PARTICIPATIVE MODEL FOR GRADUATE EDUCATION

Graduate education is a wide-spectrum effort that requires the support and participation of various groups within the academic community. Faculty, staff, and administrators should provide a supportive and conducive environment for graduate education. This requires provision of the infrastructure, administrative services and resources to facilitate graduate education and research. This is

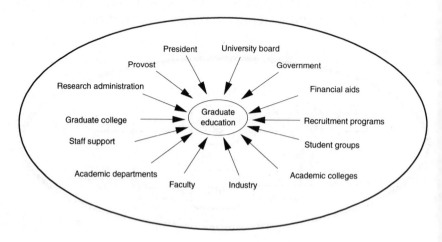

Figure 1.3 Participative model for graduate education.

particularly crucial in science and engineering programs which have special requirements.

Interdisciplinary science and engineering research programs are needed to address many of the emerging technological and environmental problems that are developing around the world. Traditionally, engineers and scientists have played major roles in attempts to solve society's problems. With the present inter-dependencies of technical, social, economic, political and natural systems, various science and engineering disciplines must come together to address the development of versatile graduate programs. The university environment must use a participative model similar to that depicted in Figure 1.3.

2 | Selecting a graduate school

Choice, not chance, determines destiny.

2.1 STATEMENT OF PURPOSE

Many graduate schools require a statement of purpose as a part of the admission application process. Write an expressive statement that highlights your strengths and preparation for the rigors of graduate study. Academic standing, professional insight, professional experience, professional maturity, communication skills and motivation are some of the factors considered in graduate admission processes.

Some pertinent information to include in your statement of purpose is:

- your understanding of the importance of your intended field of study;
- your desire and drive to pursue the study;
- ultimate career plan;
- relevance of previous work experience and academic study;
- your familiarity with research techniques and computer tools;
- your personal characteristics that support your educational objectives;
- financial resources and your need for financial aid, if applicable.

2.2 SURVEYING AVAILABLE OPTIONS

Be sure to obtain a copy of the printed handbook or guide for new graduate students from the prospective schools you are considering. Such handbooks and guides should be your frequent companions throughout your graduate studies. The questions you ask when evaluating prospective graduate schools should address the following items:

- library and research facilities;
- minimum credit hour requirements for degrees;
- thesis versus non-thesis options;
- graduate school requirements about advisors, committees and qualifying exams;
- assistantships and financial aids;
- on-campus work opportunities;
- engineering and science laboratory facilities;
- computer laboratory facilities;
- computer network facility (e.g. engineering computing network);
- document processing centers for self-service word processing;
- office space for graduate students;
- existence of relevant research centers and institutes on campus;
- job placement services;
- prestige and reputation of the school.

2.3 ADMISSION AND ENROLLMENT PROCEDURES

In order to assure efficient and timely pursuit of the graduate program of study, the student should become familiar with the prevailing administrative procedures at his or her institution. Some of the general steps in admission and enrollment are described below.

2.3.1 Admission examinations

It is recommended that the student submit scores from a recognized standardized examining board such as the Graduate Record Examination. Even where such examinations are not mandatory, it is wise to have such records on file. This may become useful for general evaluation purposes.

2.3.2 Enrollment

The student must make an application and be enrolled according to the existing graduate procedures in the institution.

2.3.3 Proficiency demonstration

You must act immediately to remove all conditions and to achieve full graduate standing. This can involve the completion of a foreign language proficiency examination, and/or demonstration of research proficiencies as may be specified by the institution or academic department. There may be a series of examinations designed to ascertain your readiness for graduate research throughout the program. The names, definitions and procedures for such examinations can vary from department to department. You should familiarize yourself with the applicable examinations. Typical examinations, when required, may be called qualifying examination, general examination or comprehensive examination.

2.3.4 Required courses

Requirements for course work at the graduate level vary greatly from institution to institution. If you are concerned about the burden of such requirements, be sure to inquire with the institution or specific department before going to all the trouble of applying for admission. This is particularly important if you have some course work deficiency from your undergraduate studies. Doctoral programs typically do not have rigid prescribed course requirements. A custom set of course work is developed for each student with the advice and guidance of the major professor. The doctoral program is developed to be consistent with the student's background with the aim of the student accomplishing the projected graduate study goals.

In some programs, there are certain core areas that all graduate students in a particular area of concentration are expected to have covered at one time or another in their academic preparations. In any case, the courses to be taken depend on several factors including the student's academic preparation, the research advisor's counsel, the graduate committee's recommendation and the specific requirements in the program. This is handled on a case-by-case and student-by-student basis.

In Europe most doctoral programs require no course work at all. The degree program consists essentially of the doctoral research, and the usual practice of having an advisory committee is not applicable. Some institutions in the USA are beginning to adopt this no-course no-committee approach to advanced degree programs. At the same time, many foreign institutions are beginning to adopt some of the features of the US system because of its flexibility. A later section discusses a European approach to graduate education.

2.4 THE ADVISORY CONFERENCE

In cases where advisory committees are used, you should request an advisory conference during the first semester of your enrollment. This is important to determine which set of courses are germane to your intended area of study. Failure to adhere to the recommendations of an advisory committee may result in loss of credits. Depending on the prevailing regulations, the advisory committee may be under no obligation to accept course work undertaken prior to the advisory conference.

The advisory conference is intended to establish your plan of study!

The membership of the advisory committee will be selected with the idea that it will become the doctoral committee. Where desirable, the advisory committee may suggest alternate faculty members for the doctoral committee. Students enrolled in a master's degree program, who are interested and have received encouragement from the faculty advisors to continue on to a doctoral program, should apply for an advisory conference as they near completion of their master's degree program. It is recommended they should have an advisory conference before enrolling for course work beyond the requirements for the master's degree because such course work must have the approval of the advisory committee. Indiscriminate enrollment without proper counseling and advice could result in taking more course work than necessary for the intended doctoral program.

Form a liaison with other students to share information about program requirements and administrative procedures. Liaison with other students has the advantage of building a spirit of collegiality and a supportive environment. In cases where courses are taken as a part of the program requirements, group investigation of research topics and ideas can also be helpful.

2.5 FAMILIARITY WITH PROGRAM REQUIREMENTS

Read and become familiar with written guidelines for your specific graduate program!

All graduate programs have published guidelines that convey the rules, regulations, and requirements for each course of study. It is very important that you become familiar with such publications. They are usually available in brief handouts that you will be given on your admission to the program or as soon as you officially enrol in the school. This has been one area of serious deficiency for a lot of students. Many students fail to familiarize themselves with the requirements and they end up pestering faculty and staff for information that is already available to them, but filed away somewhere in their homes.

Develop good work habits!

2.6 UNDERGRADUATE RESEARCH PARTICIPATION

To recruit students into graduate programs, science and engineering departments should get their undergraduates involved in research at appropriate stages in their undergraduate education. A selected group of undergraduate students can be channelled through a program that will lead them towards graduate education. The format below is suggested.

2.6.1 Goals and objectives

- Gain research experience with dedicated mentors and role models.
- Look for academic enrichment, mentoring and counseling to enhance student self-confidence, interest and desire to achieve science and engineering careers.

2.6.2 Examples of activities

- effective early faculty mentoring
- special academic and career counseling
- academic year research participation
- summer research or internship experiences
- special seminars and colloquia
- scientific conference participation
- academic enrichment activities

2.6.3 Research participation for lower-level undergraduates

Academic support

Each student should participate in a science and engineering retention program. The program can include clustering for math, science and engineering courses, academic enrichment workshops, monitoring, academic counseling and advising and early enrollment. Clustering allows for the enrollment of five to seven students in the same section of a class. The academic workshops then take these clustered students and provide extra hours of supplemental instruction, tutoring and practice tests.

Research experience

Each participant should be assigned to a faculty mentor for five hours per week. The purpose is to expose students to the world of research at their academic and developmental level and to develop a positive relationship with the faculty. The research experience should be designed for positive feedback and to encourage and enhance students' academic and personal development giving additional self-confidence and esteem.

Career planning

Students should participate in career-related activities designed to enhance their perception of career options. These activities include:

- registration and participation in on-campus interview opportunities;
- participation in interview workshops;
- resume´ writing workshops;
- participation in career fairs whenever available;
- workshops on preparing technical research papers and oral presentation.

Regular research meeting

Each student should attend periodic (e.g. monthly) meetings for participating undergraduate students. The purpose of this is to allow exchange of ideas, report on status of research activities, and provide a supportive and encouraging environment.

Student mentoring

Each lower-level undergraduate should be assigned to an upper level undergraduate who will serve as a mentor. The student-to-student mentoring arrangement facilitates near-peer relationship and avoids the possibility of an intimidating relationship between student and faculty.

2.6.4 Research participation for upper-level undergraduates

Research experience

Academically matured undergraduates can participate in the upper-level undergraduate research experience. Each upper-level student should be assigned to work on a research project for up to ten hours a week with a faculty mentor and his/her research team. The undergraduate is expected to be a contributing member of the research team.

Technical paper

Each upper-level undergraduate participant in the program should be expected to prepare a technical paper for the purpose of presentation or publication. Each will make a formal presentation of his/her research project to the rest of the group. This can be done at regular monthly meetings or specially organized seminars. The eventual objective is to prepare for presentation at national conferences.

Industrial research opportunities

The upper-level undergraduate participants will have an additional opportunity to experience research in a different environment. The institution should seek corporate sponsors that will provide summer internships in their plants and research facilities.

2.7 PARTICIPATION OF UNDERREPRESENTED STUDENTS

It is well known that women and other minority groups are seriously underrepresented in science and engineering graduate programs. The limited availability of precedents, role models and supporting resources are partly responsible for this shortage. Unfocused preparation and low level of motivation and interests on the part of the students have also been blamed. In either case, special attention should be given to needs of these underrepresented groups to enable them to participate fully in science and engineering graduate programs and research contributions. Cultural, ethnic and socio-economic diversity can greatly enhance the overall science and engineering research infrastructure.

Given the expected shortage of engineers and scientists in the coming decades, universities, corporations, government agencies and professional organizations are exploring ways to boost science and engineering enrollments and graduation rates. The increasing roles that minority groups are playing in economic development have made it imperative that these groups be targeted as potential sources of future engineers and scientists. It is crucial to increase the participation of women and minorities in engineering and science research at the graduate level. To encourage the underrepresented groups to pursue careers in engineering, new training opportunities and research guidance must be developed.

Recruitment and retention of women and minorities in science and engineering graduate programs may require innovative approaches. Recruitment efforts may involve some of the following:

- posters announcing graduate programs sent to appropriate universities;
- announcements may be placed in key publications such as *Chronicle of Higher Education*, *US Woman Engineer*, *Black Engineer*, *Minority Engineer* and other similar publications;
- recruiting video tapes sent out to prospective students;
- personal contacts made with prospective students;
- current students should be asked to encourage friends and relatives to enter graduate programs;
- information posted at career fairs at local, state, national, and international professional meetings;
- targeted scholarship programs for women and minorities in science and engineering;
- targeting women and minority students at the high school level to get them interested in science and engineering;
- encouraging undergraduates to participate in research programs that will eventually lead them to pursue advanced degree programs in science and engineering.

Once recruited into a graduate program, the retention strategies used to maintain the interest of the students and retain them in the degree program, should involve the following:

- participation and leadership roles in student and professional activities;
- involvement in tutoring programs;
- involvement in research colloquium activities;
- regular meetings with peers and colleagues in similar graduate programs.

2.8 RECIPE FOR SUCCESS

To succeed with the graduate education process, a high level of self-discipline is necessary. The items presented below offer suggestions for pursuing graduate level success:

- realize that the onus of graduate education is on the student;
- develop good work ethics;
- plan and organize for graduate education;
- explore new ideas;
- familiarize yourself with institutional requirements and guidelines;
- identify where research facilities and resources are located;
- document activities and results;
- act promptly and limit procrastinations;
- develop a plan of study as early as possible and use it as a guide;
- keep up with necessary paperwork.

2.9 EUROPEAN AND OTHER GRADUATE EDUCATION SYSTEMS

As mentioned earlier, most of Europe uses a university education format that is quite different from the American system. As specific examples, some selected systems are discussed in this section. The various systems discussed illustrate the different patterns and levels of preparation for graduate education.

2.9.1 German education system

In Germany, elementary and high school cover a period of 13 years. The 13th year is partially equivalent to the first year of college in the USA. Thus, students entering a university in Germany to study engineering or science already have very good training in the basic technical subjects.

After high school, students embark on a rigorous two-year intermediate diploma program. This program serves the purpose of weeding out those students who would not proceed to regular university degree programs. The intermediate diploma is theoretically equivalent to the bachelor's degree in the USA except that the diploma cannot be used for employment purposes. Students who are successful at the diploma level proceed directly into master's degree programs. Those who do not make the successful transition end up following some sort of technical training program that may prepare them for industrial or business employment. The two-year intermediate program is deliberately tough to facilitate the weed out process.

In one specific example, 750 students started the program and only 250 successfully moved on to the master's degree level. There are no in-semester examinations. One exam is administered at the end of each of the first two years, so there is only one chance to pass and proceed.

An alternative to the two-year intermediate diploma program is what is called a 'practical university', where students spend approximately four years. Students who do not want to or cannot enter a regular university may go to the practical university instead of embarking on the two-year intermediate diploma program. At the practical university they concentrate on vocational type training. This is comparable to the Vo-Tech system in the USA. Graduates of the practical university are employable in business or industrial settings. Thus, while the two-year intermediate diploma is theoretically equivalent to the US bachelor's degree, the practical university diploma is practically equivalent to it.

At the master's level, significant interaction is arranged with supervising professors. After the two-year intermediate diploma, it takes from three to four years to obtain the master's degree. For engineering and science students, about four courses are allowed as elective areas. Students take electives in the business area. With the recent proliferation of international exchange programs,

European students are coming to the USA, usually to take their elective courses.

Master's level students do two big research projects and one final research during their programs. The final research is equivalent to a master's thesis in the USA. One professor supervises the research work, with no graduate advisory committee structure. Academic departments are usually very small with tightly focused research areas. There are typically about three professors in each departmental research area.

One very good aspect of the German system is that the cost of higher education is so low that most people can afford it. In addition to the low cost, several government financial aids are also available. The cost is approximately US$70 per semester. This includes free transport anywhere in the city. There is typically no summer semester during the academic year.

The German example presented above is representative of what exists in many other European countries, with minor modifications here and there to suit specific national practices.

Despite the noted differences in graduate education systems around the world, the guidelines presented in this book are still applicable on a generic scale. The guidelines on dealing with graduate advisory committees can also be utilized when dealing with single supervising professors. Since most graduate research projects in Europe are sponsored by companies, the guidelines on dealing with company staff should be very useful in such settings.

2.9.2 British education system

At age 16 children take national exams in a wide spectrum of subjects. These exams used to be called O-levels. They have now been discontinued and GCSEs are the current equivalent. Students can choose to leave school after these exams or stay on for two more years and take A-levels at age 18. For those aiming to go to university, the normal number of A-level subjects is three, but students can take two or four. GCSEs are typically taken in a wide range of subjects (five to ten). Pass grades are A, B, C, D, E, with two fail grades. In order to continue to A-levels students should aim to get A, B or C grades in subjects they want to study at A-level, with a total of about five A, B or C grades overall.

Obviously students are forced to specialize after 16. They need to think about what subject they might read at university so that they can choose a suitable combination of A-level subjects. For example, a student might pass 11 O-level subjects (including English, math and French) but continue with only three subjects at A-level (math, chemistry and physics) with the intention of reading chemistry at university. As a rough rule it is a good idea to do an A-level in the same subject you would like to read at university. Traditionally, students often choose between arts and science subjects and it can prove difficult to switch back later on.

Some subjects such as economics, philosophy and geography are highly compatible with both arts and sciences. It is more popular nowadays to study sciences plus one foreign language. To do medicine or veterinary science, mathematics, physics plus chemistry or biology are needed. Alternatively, three sciences would be the normal A-level choices. Double math and physics or math, physics plus one other science would be best for physics or engineering. A-level grades are A, B, C, D, E (all passes) and O (a failure at A-level, but an O-level pass) and finally F (complete failure).

A candidate who is thinking of going to university should be aiming for two or three C or better grades. Essentially all universities in Britain are public not private universities (roughly equivalent to state universities in the USA). Several years ago, there were 33 universities and about 25 to 35 polytechnics all offering degrees (bachelor's degrees). The law changed three to four years ago and polytechnics were able to apply for university status. All of them did this and they are now universities (some have changed their names).

Admission to a university is strictly dependent on A-level grades. In other words, no matter how much money mum and dad have, if you do not get good grades in A-levels you will not get into university (even the elites have to pass some exams). At the 33 old universities most bachelor's degrees are honours degrees; a few degrees at the new universities may be ordinary degrees, which are considered inferior and roughly equivalent to passing the first two years of exams and failing the final year of an honours degree. Courses are generally three years, with some exceptions. Sandwich courses involve one year spent working in industry; courses involving a foreign language component have one year abroad. Four-year courses are slowly becoming more popular.

A student is admitted to read a specific subject in a university. This will normally not change, although there is sometimes an opportunity to switch to a closely related subject during the first year. Thereafter, you are usually absolutely stuck with your subject. This is a drastic difference from the US system, where a student can change his or her major at any time and as many times as desired.

The first year course will commonly comprise the chosen subject plus two others. A third of the time will be spent on each, with exams in each at the end of the year. You must pass these exams to be admitted to the second year. Failure means only one chance of a resit, generally at the end of the summer, just before the start of the second year. If you fail this you will be thrown off the course. Traditionally (not always), there is no continuous assessment in the British system; there are no (or very few) tests during the year. The end of year exams are therefore all important.

In the second year of the course, you would typically spend between two-thirds and all of your time on your main subject (possibly one-third of the time on a subsidiary subject). Again exams must be passed at the end of the year or you could be thrown off the course. If you are on an honours degree course, if you pass second year exams you are normally assured of an ordinary degree even if you fail your final year exams. Final exams at the end of the third and final year are all important, typically counting for more than 50% of the degree marks (the exact percentage will vary from course to course and university to university). Degrees are usually BSc (Hons) or BA (Hons) – bachelor of science, bachelor of arts. Grades are usually first, second (usually divided into 2(i) and 2(ii)), third, ordinary and fail. On a typical course, most students will get seconds (perhaps 55–60%). Thirds will probably have the next highest percentage, with only 5–10% getting firsts. Ordinary passes and fails (dropouts, etc.) will together make up about 10%.

Students who obtain a first or 2(i) bachelor's degree can choose to study for a PhD. It is not necessary for these students to get a master's degree first. Students can stay at the same university or change. Grants are awarded to university departments by government funding agencies (e.g. Engineering and Physical Science Research Council (EPSRC) and Medical Research Council (MRC)) according to their prowess (grading) in research in that discipline. Some PhD awards are also jointly funded by these government agencies and industry.

Once a department is awarded a PhD place, it is free to choose a candidate (normally a student with a first or 2(i)). The grant will be for three years and covers equipment expenses and a small living allowance. The student will aim to finish all experiments and research work within the three years, and if possible to submit the thesis as well. These days there is pressure for students to submit their thesis reports before four years are up.

A PhD is all research, with no lectures or course work. PhDs are also nearly always pass/fail (no letter grades). A master's degree, in addition to being a qualification in its own right, is a route into a PhD for someone who has a lower bachelor's grade. Usually lasting one year, a master's degree program typically comprises lectures and short research projects with exams and a mini thesis at the end of the year.

2.9.3 The Greek system

In the Greek education system, there are six years of elementary school and three years of high school, making up a total of 12 grades. Up to and including the 11th grade, the curriculum is uniformly preset for general education by the Ministry of Education and Religion. For the 12th grade, students choose their areas of study (major) and, accordingly, the type of school to attend. Typically, there are four areas:

- engineering and science
- medically related professions
- social sciences and humanities
- business and economics

In addition, there are three types of studies, called 'specialization', which students enter and choose as major after the 9th grade:

- classical (art, literature, social sciences, humanities)
- science
- vocational

The first two have entrance exams. Upon the completion of all high school requirements, students may apply to universities, except for the students from vocational schools who usually only apply to lower-level colleges. Each university has four entrance exams, the subjects depending on the previously chosen major.

Undergraduate programs typically last four years, except for engineering and architecture which are five years, and medicine, which requires eight years. The grading scale is 5 to 10, with 6 being the lowest passing grade. There is usually only one (final) exam for each course at the end of the semester. Students who fail a course need not attend the lectures again. They only need to retake the exam offered six to nine months later.

Graduate master's and PhD programs are usually very specific in an area and are almost exclusively research oriented (few or no courses). There are only three universities that offer master's programs based on heavy course work: examples are

- international politics and relations at the University of PADIOS in Athens;
- telecommunications at the University of Athens;
- history of Macedonian studies at the University of Thessaloniki.

In general, as in the USA, each student does research and writes a thesis which must be approved and later presented to a three-member committee. Some of the committee members may be from a different university. Some graduate programs require the students to attend at least one semester abroad, at an approved university.

2.9.4 Graduate education in Russia

Typically it takes four to five years to get a first degree. The first three years roughly correspond to US bachelor's programs. Students take only the required courses. The third and the fourth years are more like a graduate school. Students can take elective courses and conduct research under the supervision of an advisor (with no graduate committee). At the end of the fifth year, students take a state examination in their specialism and present a thesis (diploma work). After a successful completion, a diploma is awarded to the student. This is equivalent to a US master's degree.

No undergraduate degrees are awarded. Those who want to continue their education take exams and after successful completion, are admitted to a doctoral program. It takes three to four years to write a dissertation. Typically, doctoral students do not take any classes; they just undertake research. Again, there are no graduate committees, just a dissertation research advisor. After successful presentation of the research, a candidate is awarded a degree. This is equivalent to a PhD in the USA.

At this stage, all the paperwork on the research is sent to a state committee appointed by the Ministry of Higher Education. This state committee approves the degree. Thus, a standard is established to make sure that a degree obtained from a smaller university is at the same level with degrees obtained from the best universities. There is often a danger of the state committee abusing its powers in granting doctoral degrees. The systems in most of Eastern Europe (e.g. Bulgaria) are similar to the Russian system. These systems are currently undergoing changes in many of the countries.

2.9.6 The Chinese system

Undergraduate education in China takes four years, after which a bachelor's degree is awarded. To enter a master's program, students take highly competitive exams. A master's degree can be obtained in three years. The student must write a master's thesis. A PhD program takes another three years. Only senior professors are allowed to have PhD students. There are thus very few PhDs emerging from those programs. Most students aspiring to doctoral degrees go to the USA for that purpose.

2.9.7 The Saudi system

In Saudi Arabia, elementary school, intermediate school and high school cover a period of 12 years. When a student finishes the intermediate school, he or she is given the chance to select from four different high school programs. These high school programs are regular high school, modern high school, industrial high school and trade high school. The first two high school programs qualify the student to apply to university and enroll in an appropriate college.

The regular high school program follows the British system, and the modern high school program follows the American system. In the regular high school program, all students study the general curriculum for the first year. Then, for the two remaining years, the students can select either a liberal arts major or a science major. The modern high school program gives students full flexibility to select and plan their majors from the first year. The student is given the chance to select from four different majors. These majors are Islamic and literary studies, humanitarian and administrative science and natural science (physics and mathematics or chemistry and biology). The industrial high school program qualifies the student to either work in industrial fields or apply to technology college, which gives a two-year diploma in advanced practical training. The trade high school program qualifies the student to either work in administrative fields or apply to business and administration college in the university.

There are seven universities in Saudi Arabia. Four of the universities offer all majors and follow the modified American system (MAS). The other three universities which teach mainly liberal arts subscribe to the British system. The MAS follows the credit system which is applied in the USA. In the MAS, however, the students are not given the chance to select their courses; the courses are chosen for them. In addition students can add and drop to switch between sections, but they are not allowed to reduce their load. In the British system, a group of students will start a specific major and will finish their degrees together except for those who fail the program.

The same two options are available to students who want to complete their graduate studies. The master's programs in the MAS require the students to take some courses and then start working on their theses. In the British system, no courses are required and the students will start their research in the first semester. Most students proceed abroad for doctorate level studies.

Selecting a research topic | 3

'The mere formulation of a problem is far more often essential than its solution, which may be merely a matter of mathematical or experimental skill. To raise new questions, new possibilities, to regard old problems from a new angle requires creative imagination and marks real advances in science.' Albert Einstein

3.1 RESEARCH DEFINITION

Research = Developing a new idea and proving that it works

For the purpose of graduate education, research can be defined as developing a new idea and proving that the idea works. This requires very careful research formulation strategy. A feasible and compatible research topic must be selected for the research to be successful.

- Does the topic fit your background and academic preparation?
- How well does the topic fit your personal and career interests?
- Is the topic likely to lead to a new contribution to the field?

A common problem in evaluating graduate research is determining what new work has been done by the student and what has been culled from the existing literature. To address this problem, a detailed write up must be presented on the proposed methodology.

One source of motivation for a research topic is a real-world work environment. Graduate students who have been in the real-world work environment often have a good idea on what their research topic should be. Formulate your overall research goal in terms of achievable objectives. Each research effort faces three major constraints:

- performance expectations
- schedule requirements
- cost limitations

To address the prevailing constraints adequately, you must be ready for a paradigm shift, if necessary, in your research endeavors. A paradigm is a model or framework within which a problem is solved. It defines the boundaries and rules that guide the problem-solving approach. Paradigm shift requires looking at the other side of the coin, so to speak. The idea that what was successful in the past will continue to be successful does not pave the way for research success. Figure 3.1 presents a model for successful graduate research management.

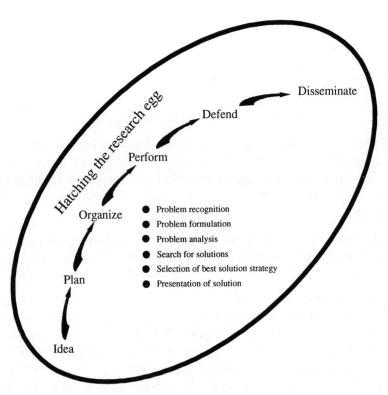

Figure 3.1 Stages of successful graduate research.

3.2 TYPES OF RESEARCH

The word research comes from a prefix and a root word meaning to seek out again. Most graduate research involves seeking out ideas and materials already developed or documented by others. In this process, discoveries are made. Some research endeavors, particularly in science, focus on finding out things that have never before been known or documented. In this process, new inventions may develop. Different types of research are described below:

- **Laboratory research**: research performed in a laboratory setting. This may use live test subjects (the proverbial guinea pigs), or inanimate laboratory materials.
- **Experimental research**: research performed by setting up carefully designed experimental procedures. This involves studying the effects of certain combinations of independent factors (treatments) on some dependent factor. Most laboratory research studies are experimental.
- **Pure research**: research done purely for scientific interest and curiosity. It aims to add to the body of scientific knowledge. It often deals with physical and scientific objects and phenomena. It may not have immediate or practical application potential.
- **Scholarly research**: research effort where the research deals primarily with written documents rather than physical and scientific objects, as in pure research. The publication of papers in technical journals and presentation of technical papers at professional conferences fall under the category of scholarly research.
- **Applied research**: research that focuses on practical applications of what has already been discovered, developed, documented or theorized.
- **Theoretical research**: research that addresses the development of new concepts based on proven scientific merit. It adds to the body of knowledge from which practical applications can later be drawn.
- **Technical research**: research involving the study or development of physical objects with the intended purpose of practical utility. It may address the functionality of the object or the scientific merit of its configuration.
- **Business research**: research dealing with information needed to make business-oriented decisions. It often involves the development of strategic business plans.

- **Economic research**: research that addresses the processes of interactions of economic factors and their implications for business, social and political events.
- **Market research**: research involving the study of what the market (consumers) wants. This helps determine the type, nature and quantity of products to develop for the consumers.

3.3 METHODOLOGIES AND STRATEGIES

'Everything should be made as simple as possible, not one bit simpler.' Albert Einstein

For any type of research, you need to develop good methodologies and strategies. Write a concise objective statement and review it frequently (e.g. once a week) to avoid irrelevant digression from the focus of the research. Table 3.1 presents an outline for research performance. The overall layout of your research methodology might look like this:

Abstract
Introduction
Methodology
Research protocol
Experimental procedure
Results
Conclusion

Table 3.1 Outline for research methodology

Element	Description
Research hypothesis	Specify the research question to be addressed (what, why)
Define problem scenario	Explain the situation relevant to the problem (who, what, where)
Research goal	Specify what is to be done
Research plan	Develop the plan to carry out the research (who, what, when, where, how)
Implementation	Carry out the research
Conclusion	Draw conclusions from the results
Documentation	Report the findings

Key questions for prospective research are:

- What do you intend to do?
- Why is the work important?
- What has already been done?
- How are you going to do the work?

To address these questions, organize your strategy as follows:

1. Clear problem statement:
    (a) start with an open mind;
    (b) don't jump to conclusions;
    (c) identify problem to be solved;
    (d) develop your hypotheses;
    (e) define what you want to do;
    (f) decide what you can accomplish.
2. Significance of proposed work:
    (a) problem background;
    (b) relevant literature;
    (c) gaps that need to be filled.
3. Justification for the problem:
    (a) the discipline and scope;
    (b) fields relevant to the research;
    (c) intrinsic merit.
4. Feasibility of research:
    (a) validity of approach;
    (b) qualification and preparation for the research;
    (c) available resources;
    (d) preliminary study.
5. Experimental plan:
    (a) project organization;
    (b) innovative aspect of the methodology;
    (c) feasibility, adequacy and appropriateness of the approach;
    (d) difficulties anticipated;
    (e) contingency approaches;
    (f) timeline (activities and schedule).
6. Research evaluation:
    (a) data analysis;
    (b) interpretation of results;
    (c) milestones.
7. Follow-up:
    (a) review the obvious;
    (b) continuation of research;

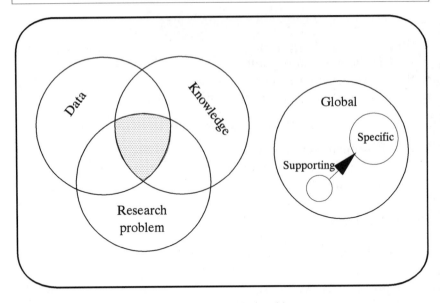

Figure 3.2 Problem, data and knowledge relationships.

(c) long range impact;
(d) dissemination.

Figure 3.2 illustrates the relationships between problem, data and knowledge in science and engineering graduate research. To have a feasible research goal, a workable intersection must exist between the available data, the existing knowledge and the problem definition. Within the global problem structure, a specific focused research problem must be identified. In the research work, it may be necessary to identify supporting areas that lend data and knowledge to the research at hand.

3.4 HOW TO CONDUCT A LIBRARY SEARCH

A successful library search is the first step towards successful graduate research. Library research for science and engineering graduate students often takes place in specialized library facilities, such as an engineering library, where materials in the engineering field and the related fields are the focus of the collection. As in other general libraries, materials are arranged by call numbers in open stacks. (Call numbers enable you to locate a book on the shelf.) According to the Library of Congress (LC) classification system, the call number begins with one, two or three letters plus an arabic number, followed by the Cutter number, which is the code of either the author, or the title of the book. Sometimes, there is a four-digit number at the very end, which is the date of the publication of the book (e.g. QA76.C14 1993).

3.4.1 How to find books

A computerized catalog in an online system helps students locate books fast and easily. The catalog usually provides access by author, title, subject, keyword and call number. For author searches, the author's name must be inverted (you must input the last name before the first name). For title searches, the first word must be omitted if it is *a, an* or *the* (e.g. theory of relativity, not *the* theory of relativity). For subject searches, the appropriate Library of Congress subject heading must be used (subject headings may be determined by using the Library of Congress subject heading volumes).

The advantage of subject searches is that you can locate many books related to your subject(s). But the disadvantage is that it is difficult to go through the hassle of finding the right subject heading in the three-volume publication. One way of taking care of this problem is to go to the shelf and browse after finding one book. You will usually find many subject-related books in the same area.

Another method is a keyword search. Keyword searches allow use of everyday language and combination of terms. Because they are easy, fast and effective, most students prefer keyword searches to other approaches. The computerized catalog also provides access to periodicals, government documents, non-book materials (e.g. microfiche, microfilm, audio–video, etc.), theses and dissertations.

3.4.2 How to find articles

Indexing and abstracting services are the primary access to articles in periodicals and proceedings. In the engineering field, as well as the most often used index, *Engineering Index*, many other specialized indexes are also important tools. Some libraries have them on CD-ROM, some are paper bound, depending on the library's budget situation.

In most indexes the citations can be found by looking up a subject or author. Citations will generally include the name of the author, title of the article, title (or abbreviation) of the publication in which the article appears, identification of the appropriate issue (volume number, issue number, date) and page numbers. Abstracting services also include a short summary of the article.

3.4.3 Library services

Interlibrary loan services are available for materials not found in local engineering libraries. It generally takes about two or three weeks to receive books or photocopies of articles through interlibrary loan. Lending libraries set rules and restrictions for interlibrary loan materials, including the loan period, renewal options, and cost. Most of the time, these services are free. These services are very much welcomed and heavily used by students and faculty. Locations of lender(s) can be found through RLIN, OCLC, Internet, etc.

Library orientations are usually scheduled at the beginning of each term and upon request. Library instruction sessions for classes may be arranged by instructor. One-on-one orientation is also available depending on the situation.

3.4.4 Library tips

Always ask the reference librarians for assistance. Because they are trained, qualified, service-oriented professionals with many years of experience, they are always ready and willing to answer different types of research question. If you want to find a short cut to save time in your research, the reference librarian is the right person to turn to for assistance.

3.4.5 Finding biographical information

BIO-BASE is an index to biographical material found in over 600 dictionaries, encyclopedias and *Who's Whos*. If you need biographical information about a person, *BIO-BASE* allows you to determine quickly in which books to find it. It can also refer you to other indexes which will give you further leads to books, magazines, and newspapers. One such index is *Biography Index*, which works like the reader's guide.

In *BIO-BASE*, names are listed on microfiche in alphabetical order. To check for listings on a person, pick the appropriate envelope. The entries give last name, first name, middle name, birth date (death date), and abbreviations for the source books. The *BIO-BASE* manual gives an alphabetical list of these abbreviations and what they stand for. For example:

- BioIn 11 stands for *Biography Index*, Volume 11
- CngDr 85 stands for *Congressional Directory* (1985)
- WhoAmW 87 stands for *Who's Who of American Women* (1987)

Call numbers are written next to the abbreviations for books in the library's collection. If a given person is not listed in *BIO-BASE* under any name, it means that no biographical material on that person can be found in any of the many sources indexed by *BIO-BASE*. For people that are hard to find, ask a librarian for help.

3.4.6 Literature review summary

When a literature search is done on a subject, two things can happen. You will either find very limited published materials on the subject or you will encounter an overwhelmingly large number of relevant materials. You must learn to prune the literature and sort out the items that are really useful to your research. Once you narrow down the materials, you should briefly summarize each one. This is useful for cross-checking methodologies later on and writing the literature review section of the technical report. The example format presented below is useful for that purpose. Summarize each relevant literature find as follows:

Title: A new computational search technique for AI based on Cantor set
Author: Adedeji B. Badiru
Journal: *Applied Mathematics and Computation*, Vol. 57, 1993, pp. 255–274.
Abstract: This paper discusses the development of a new search algorithm for artificial intelligence systems. The search technique is based on the mathematical theory of Cantor set. The search algorithm will be efficient for specialized search domains where the distribution of the data elements to be searched is approximately normal. The approach uses the iterative procedure of deleted middle thirds. This facilitates

quick pruning of a search space. The new search technique has potential applications in computational systems with large input–output data handling. Extensive experiments comparing the Cantor set search (CSS) to binary search indicate that the search technique holds good promise.

Solution approach: The paper uses a methodology based on the theory of Cantor set. Search distribution is limited to bell-shaped curves.

Computational experience: Comparison to binary search.

Conclusions: The new search technique holds promise based on comparative results with binary search. Further research and extension to other types of distributions is needed. The computational complexity of the Cantor set search algorithm is yet to be developed.

3.5 THE RESEARCH PROPOSAL

Even the best idea is worthless if it cannot be sold to anyone.

If a research proposal is done well, the job of writing the research report later on will be simplified. For most research efforts, it is effective to document results as the research progresses. In the proposal, you should:

- develop a narrow, focused thesis statement;
- be able to say in one sentence what your research objectives are;
- be able to say in one sentence what your research methodology is;
- be able to say in one sentence what your contribution is.

If you have difficulty with this you need to seriously rethink your research agenda. A basic requirement of successful research is being able to state your research concisely and convince the audience that it is a worthwhile effort. In addition to definitional discussions and narrative on the problem area, you need a 'rigorous' section on literature surveys to demonstrate familiarity with other researchers' works. The outline of your research proposal should look something like the following:

I. Introduction
   Background

Problem statement
Objectives
Merit or utility of the research
II. Proposed methodology
Literature survey
Relevance to previous works
Unique aspects of the research
What is new
III. Expected contribution
Implementation requirements
Limitations (if any)
IV. Conclusions
V. References

Questions to be prepared for in defending your research plan include the following:

- What is new in your research?
- What contribution to the literature will your research make?
- Where and how can your research results be implemented?

3.5.1 Approval of topic

To get approval from your advisor for your proposed research topic, develop a 'topic justification' outline as presented below. Make it concise enough to fit within no more than two pages. This outline is not a formal research proposal, but a brief indication of the justification for the research.

- **Topic**: specify topic area or title.
- **Problem area**: identify the problem area to focus on.
- **Importance of the problem**: describe the implications of the problem.
- **Existing or conventional approach**: describe method of solving the problem.
- **Shortcomings of existing approach**: explain drawbacks of existing method.
- **Proposed methodology**: outline your proposed approach.
- **Validation**: describe your validation strategy if applicable.
- **Expected results**: identify what is expected from the research.
- **Criterion measure**: identify the performance measures for the research.
- **Contribution**: establish the contribution of the research.

3.6 PROPOSAL EVALUATION CHECKLIST

On a scale of 1 to 5 (1 = poorest, 5 = best), rate each component listed below for the various elements of the research proposal. Not all the elements included below will be applicable to all proposals. It may be necessary to condense or expand the list based on specific situations and interests.

Abstract/research summary	*1*	*2*	*3*	*4*	*5*	*Comments*
1. Is preceded by title						
2. Appears at beginning of proposal						
3. Identifies the student's area of study, department, etc.						
4. Includes at least one sentence on problem statement						
5. Includes at least one sentence on objectives						
6. Includes at least one sentence on methodology						
7. Is brief and concise						
8. Is clear						
9. Is relevant to the area of study						

Statement of problem	*1*	*2*	*3*	*4*	*5*	*Comments*
1. Concisely states what is to be done						
2. Is of reasonable scope						
3. Is supported by literature and previous related works						
4. Indicates new idea, creativity, or innovativeness						
5. Gives evidence of knowledge of the field						
6. Is interesting and timely						
7. Is feasible and shows potential for success						
8. Makes no unfounded assumptions						

Research objectives	*1*	*2*	*3*	*4*	*5*	*Comments*
1. Relates to the statement of the problem						
2. Makes it clear that the objectives represent expected outcomes						
3. Makes it clear that objectives are not methods						
4. Describes expected benefits						
5. Identifies the performance measure or criterion measure						

The methodology	*1*	*2*	*3*	*4*	*5*	*Comments*
1. Flows naturally from problem statement and objectives						
2. Clearly describes experimental or research activities						
3. Shows evidence of good scientific principles						
4. Describes justification for the proposed approach						
5. Describes how data will be collected, handled, and analyzed						
6. Describes the research facilities and tools to be used						
7. Outlines scope and sequence of activities within the time frame and resources available						

3.7 INDUSTRY-SPONSORED RESEARCH

Graduate research sponsored by a profit-oriented company deserves special attention. The company must be convinced of the importance and benefit of the research. The presentation of such research results must focus on its impact on the company's 'bottom-line'. Industry-sponsored projects can cover three primary objectives:

1. To identify current methods, approaches and organizational linkages used to implement manufacturing/industrial extension education programs in the universities.
2. To assess the universities' current capacity, available resources and expertise, and willingness/ability to meet future require-ments for an effective industrial extension program.
3. To identify appropriate intra-university and inter-company linkages vital to the implementation of university–industry collaborative research.

For industry-sponsored projects, provide answers to the questions in the following subsections.

3.7.1 The problem

1. What exactly is the problem?
2. How important is the problem to company operations?
3. What specific evidence of the problem is available?

4. What has already been done about the problem?
5. What happens if the problem is not addressed?

3.7.2 The methodology

1. What are the expected benefits of the proposed solution?
2. What are the anticipated difficulties, limitations and requirements?
3. What are the risks associated with the solution proposal?
4. What contingency plans can be made for the research?
5. How is the proposed methodology justified over other approaches?

3.7.3 The implementation

1. When can the proposed solution be implemented?
2. How long will the implementation take?
3. What resources, costs and actions are needed for the implementation?
4. What are the success metrics (evaluation approach) for the implementation?
5. What are the milestones to measure progress of the implementation?

3.7.4 The benefit

1. What are the short- and long-term impacts of the solution?
2. What is the return on investment for the solution?
3. What is the benefit–cost ratio for the solution?
4. What guarantees (if any) are available for the expected returns?
5. What are the benefits for future research collaborations?

In dealing with company staff, it is quite possible that different levels of management will be encountered. Presentations should be tailored to the interests of the different groups. Table 3.2 presents items to focus on for selected groups of company staff.

Table 3.2 Contents of presentations to company staff

Senior management	Middle management	Technical staff
Return on investment	Increased production	Research and development benefit
Strategic planning	Reduced personnel problems	Introduction of new technology
Increased market share	Better customer service	Technical detail of methodology
Improved decision tools	Employee satisfaction	Project schedule
Long-term growth	Cost control	Computer interface
Capital improvement	Better operational planning	Training requirements
Cost reduction	Production schedules	Soundness of technical approach
Community relation	Information access and management	Interface with company process

3.8 ASSESSING UNIVERSITY CAPABILITY FOR INDUSTRIAL COLLABORATION

To assess current capability and existing resources for industrial collaboration, consider the questions presented in this section. The expected contribution of a university in industrial collaboration is to help a company improve its products, services, processes and, consequently, profits. Bilaterally, the expected contribution of a company is to provide funding and a practical testbed for university research. In industry, the idea of 'zero defects' makes sense. But in academia, 'zero defects' makes no sense since we cannot guarantee the research success of each student, try as we may. This notion can affect the expectation of industry concerning university expertise and products.

3.8.1 Industry (characteristic keywords)

- profit driven
- competition conscious
- cost awareness
- practicality focus

3.8.2 Academia (characteristic keywords)

- knowledge oriented
- exploration driven
- constrained time environment (academic year cycle)
- individualistic research focus

Answers to the following questions will shed light on the respective capabilities of the two seemingly incompatible research collaborators.

1. What specific industrial problems will the joint effort focus on? What is the cause of the problem? For example, regulatory requirement, new manufacturing process, infrastructural development, use of new technology, or plant, equipment and training needs may call for a company-sponsored research.

2. How is the need for the collaborative research identified? Did the company approach the university? Did the university solicit the project?

3. What is the stimulus for university involvement? Is is initiated by an individual faculty member, an organized program (e.g. a research center) or a need to find an application for a research project?

4. What can actually be done in the collaborative project? Examples might include combination of specific technology, process, systems applications, management consultancy, worker training, education program, plant layout, cost analysis or evaluation of alternative technologies.

5. How will the university arrange to address the industry problem? Is there an existing organizational structure (e.g. a laboratory) within which the project can be carried out? Is there an individual or faculty team to manage the interface between the university and the industry? Is there an interdisciplinary arrangement to tap into the university expertise from relevant fields?

6. What university resources, expertise, departments or disciplines will be involved in the university–industry collaborative effort? Will the university be reimbursed for any of the associated costs of using these resources?

7. What external resources and expertise will be needed for the collaborative project? How will the necessary linkages be created and maintained?

8. What are the expected impacts of the project (e.g. jobs, profits, new products, better quality, cost reduction, etc.)? What training opportunities will be available to students as a result of this project?
9. How long is the project expected to last? Can university resources and company resources be tied up for that long?
10. Does the university have the interest and willingness to commit additional resources and initiate organizational requirements to pursue future industrial collaboration?

4 | Graduate research planning

A plan is the map of the wise.

4.1 TRIPLE C MODEL

The Triple C model (Badiru and Pulat, 1995) is an effective project planning tool that can be adopted for graduate research management. The model states that project management can be enhanced by implementing it within the integrated functions of:

- communication
- cooperation
- coordination

The model facilitates a systematic approach to project planning, organizing, scheduling, and control. The Triple C suggests communication as the first and foremost function. It highlights what must be done and when. It can also help to identify the resources (manpower, equipment, laboratory facilities, etc.) required for each effort. It points out important questions such as:

- Does each project participant know what the objective is?
- Does each participant know his or her role in achieving the objective?
- What obstacles may prevent a participant from playing his or her role effectively?

4.2 RESEARCH COMMUNICATION

Communication makes working together possible. The communication involves making all those concerned aware of project requirements and progress such as the following:

- the scope of the project;
- the personnel contribution required;
- the expected cost and merits of the project;
- the project organization and implementation plan;
- the potential adverse effects if the project should fail;
- the alternatives, if any, for achieving the project goal;
- the potential direct and indirect benefits of the project.

Under the Triple C model, research communication may be carried out in one or more of the following formats:

- one to many;
- one to one;
- many to one;
- written and formal;
- written and informal;
- oral and formal;
- oral and informal.

4.3 RESEARCH COOPERATION

Cooperation is a basic virtue of human interaction. More projects fail due to a lack of cooperation and commitment than any other project factors. To secure and retain the cooperation of project participants, their first reaction to the project must be positive. The most positive aspects of a project should be the first items of project communication. For research management, different types of cooperation that should be addressed are:

- technical
- functional
- resource
- social
- administrative
- proximity
- safety

Whichever type of cooperation is available in a research environment, the cooperative forces should be channeled toward achieving research goals. Clarification of project priorities will facilitate personnel cooperation. Relative priorities of multiple projects should be specified so that a project that is of high priority to one member of the research team will also be of high priority to all team members. Some guidelines for securing cooperation are:

- establish achievable goals for the project;
- clearly outline the individual commitments required;
- integrate project priorities with existing priorities;
- eliminate the fear of job loss due to industrialization;
- anticipate and eliminate potential sources of conflict;
- use an open-door policy to address project grievances;
- remove skepticism by documenting the merits of the project.

Cooperation must be supported with commitment. To cooperate is to support the ideas of a project. To commit is to willingly and actively participate in project efforts again and again through the thick and thin of the project.

4.4 RESEARCH COORDINATION

After successfully initiating the communication and cooperation functions, the efforts of the research team must be coordinated. Coordination facilitates harmonious organization of activities. The

Table 4.1 Sample layout of research responsibility matrix

Responsibility/task codes:	Person responsible							Status of task					
	PROFA	PROFB	STUDENT	LAB	DEPT	UNIV	OTHERS	JAN 31	FEB 15	FEB 28	MAR 08	MAR 15	MAR 21
R = responsible													
I = inform													
C = consult													
S = support													
O = on track (task on schedule													
D = done (task is completed)													
B = delayed (task is late)													
Tasks													
1. Graduate admission						R		D					
2. Research topic	R	R	R	I	I				O				
3. Research proposal	C	C	R	I					O				
4. Plan of study	R	R	R			I		D					
5. Qualifying exam	R		R		I	I							
6. Experiment			R	S									
7. Data analysis	I	I	R	S									
8. Report	I	I	R										
9. Defend	C	C	R		I	I							
10. Publish	R		R										

construction of a responsibility chart can be very helpful at this stage. This is a matrix consisting of columns of individual or functional departments and rows of required actions. Cells within the matrix are filled with relationship codes that indicate who is responsible for what. Table 4.1 illustrates the layout of a responsibility matrix. The matrix helps to avoid neglecting crucial communication requirements and research obligations. It can help resolve questions such as:

- Who is to do what?
- How long will it take?
- Who is to inform whom of what?
- Whose approval is needed for what?
- Who is responsible for which results?
- What personnel interfaces are required?
- What support is needed from whom and when?

4.5 INDUSTRIAL RESEARCH COLLABORATION

Many graduate students, particularly at the PhD levels are industrially funded or are engaged in industry-driven research projects. Industrially based PhD research projects offer great opportunities to students, but also present difficulties. The difficulties are frequently in terms of discipline and mismatch of university–industry needs and cultures. The difficulties can be overcome with proper research management strategies. Figure 4.1 shows a potential model for university–industry interface for graduate research. The model provides a feedback loop process that facilitates continuing positive interaction between university and industry.

Follow the guidelines below to facilitate cooperation with industrial sponsors:

- get acquainted with an interested liaison in the company;
- formulate research scope with the goals of the company;
- emphasize the expected benefits to the company;
- maintain frequent contacts with the liaison;
- do not profess to have all the answers to the company problems;
- listen to and learn from the company's personnel.

As a part of your research management strategy, you are encouraged to work on extra projects, even if unfunded, to give you additional research exposure and experience. Summer internship

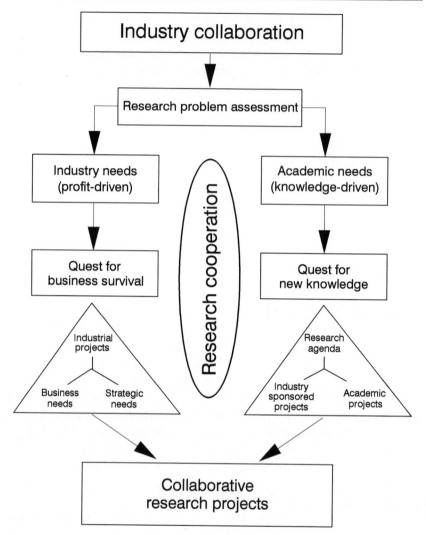

Figure 4.1 University–industry research interaction model.

in industry is one very good approach. These complementary undertakings will enhance your research competence.

4.6 INTERFACE OF RESEARCH AND PRACTICE

Much archival research is being done in various areas of engineering and science. But few of the research results make the transition from research to practice. One reason for this is that most of these results are published in the types of journal that are rarely read by practitioners. Meanwhile practice in a particular industry may continue to recycle the same age-old concepts that may no longer be valid in modern environments. It is important to institute a marriage of traditional practice and modern research. Such coupling will facilitate more practical, effective and timely tools for engineering and science. The university–industry model presented in Figure 4.1 can facilitate the coupling of academic research and industrial practice. The profit-driven objectives of practice should be integrated with the knowledge-driven goals of research.

4.7 GUIDE TO INDUSTRIAL RESEARCH PROPOSALS

Proposals are classified as either 'solicited' or 'unsolicited'. Solicited proposals are those written in response to a request for proposal (RFP), while unsolicited ones are those written without a formal invitation from the funding source. Many proposals are written under competitive bids. If an RFP is issued, it should include statements about project scope, funding level, performance criteria and deadlines.

The purpose of the RFP is to identify companies that are qualified to conduct the project successfully in a cost-effective manner. Formal RFPs are sometimes issued to only a selected list of bidders who have previously been evaluated as being qualified. These may be referred to as targeted RFPs. In some cases, general or open RFPs are issued and whoever is interested may bid for the project. This, however, has been found to be inefficient in many respects. Ambitious, but unqualified, organizations waste valuable time preparing losing proposals. The receiving agency, on the other hand, spends much time reviewing and rejecting worthless proposals. Open proposals do, however, have proponents who praise their 'equal opportunity' approach.

In industry, each organization has its own RFP format, content and procedures. The request is called by different names including PI (procurement invitation), PR (procurement request), RFB (request for bid) or IFB (invitation for bids). In some countries, this is sometimes referred to as request for tender (RFT). Irrespective of the format used, an RFP should request information on the bidder's costs, technical capability, management and other characteristics. It should, in turn, furnish sufficient information on the expected work. A typical detailed RFP should include:

1. **Project background**: need, scope, preliminary studies and results.
2. **Project deliverables and deadlines**: what products are expected from the project, when the products are expected and how the products will be delivered.
3. **Project performance specifications**: it may sometimes be more advisable to specify system requirements rather than rigid specifications. This gives the system or project analysts the flexibility to utilize the most updated and cost effective technology in meeting the requirements. If rigid specifications are given, what is specified is what will be provided regardless of the cost and the level of efficiency.
4. **Funding level**: this is sometimes not specified because of non-disclosure policies or because of budget uncertainties. However, whenever possible, the funding level should be indicated in the RFP.
5. **Reporting requirements**: project reviews, format, number and frequency of written reports, oral communication, financial disclosure and other requirements should be specified.
6. **Contract administration**: guidelines for data management, proprietary work, progress monitoring, proposal evaluation procedure, requirements for inventions, trade secrets, copyrights and so on should be included in the RFP.
7. **Special requirements (as applicable)**: facility access restrictions, equal opportunity/affirmative actions, small business support, access facilities for the handicapped, false statement penalties, cost sharing, compliance with government regulations and so on should be included if applicable.
8. **Boilerplates (as applicable)**: these are special requirements that specify the specific ways certain project items are handled. Boilerplates are usually written on the basis of organizational policy and are not normally subject to conditional changes. For example, an organization may have a policy that requires that no more than 50% of a contract award be paid prior to the completion of the contract. Boilerplates are quite common in government-related projects. Thus, large projects may need boilerplates dealing with environmental impacts, social contribution and financial requirements.

4.7.1 Proposal preparation

Whether responding to an RFP or preparing an unsolicited proposal, care must be taken to provide enough detail to permit an accurate assessment of a project proposal. The proposing organization will need to find out the following:

- project time-frame;
- level of competition;
- the agency's available budget;
- the organization of the agency;
- the person to contact within the agency;
- previous contracts awarded by the agency;
- exact procedures used in awarding contracts;
- nature of the work done by the funding agency.

The proposal should present the detailed plan for executing the proposed project. The proposal may be directed to a management team within the same organization or to an external organization. However, the same level of professional preparation should be practiced for both internal and external proposals. The proposal contents may be written in two parts: a technical section and a management section.

Technical section of project proposal

1. Project background:
   (a) expertise in the project area;
   (b) project scope;
   (c) primary objectives;
   (d) secondary objectives.
2. Technical approach:
   (a) required technology;
   (b) available technology;
   (c) problems and their resolutions;
   (d) work breakdown structure.
3. Work statement:
   (a) task definitions and list;
   (b) expectations.
4. Schedule:
   (a) Gantt charts;
   (b) milestones;

    (c) deadlines.
5. Project deliverables.
6. The value of the project:
    (a) significance;
    (b) benefit;
    (c) impact.

Management section of project proposal

1. Project staff and experience:
    (a) staff vita.
2. Organization:
    (a) task assignment;
    (b) project manager, liaison, assistants, consultants, etc.
3. Cost analysis:
    (a) personnel cost;
    (b) equipment and materials;
    (c) computing cost;
    (d) travel;
    (e) documentation preparation;
    (f) cost sharing;
    (g) facilities cost.
4. Delivery dates:
    (a) specified deliverables.
5. Quality control measures:
    (a) rework policy.
6. Progress and performance monitoring:
    (a) productivity measurement.
7. Cost control measures.

An executive summary or cover letter may accompany the proposal. The summary should briefly state the capability of the proposing organization in terms of previous experience on similar projects; the unique qualifications of the project personnel; the advantages of the organization over other potential bidders; and the reasons why the project should be awarded to the bidder.

4.7.2 Legal considerations

Research liability laws vary in different parts of the world. Nowadays, the legal aspects of research projects are more difficult and complicated than in the past. The work force is more volatile, technology is more dynamic, and society is less predictable. Governmental, institutional and personnel changes are some of the factors that can complicate the legal requirements in a research environment. The number of legal issues that can arise is increasing at an alarming rate. Any prudent researcher of today should give serious consideration to the legal implications of research undertakings. Many organizations that have failed to recognize the legal consequences of new uncharted research efforts have paid dearly for their mistakes.

With the emergence of new technology and complex systems (e.g. genetic engineering, silicon breast implants and so on) it is only prudent to anticipate dangerous and unmanageable events. The key to preventing disasters is thoughtful planning and cautious research management. Industrial research projects are particularly prone to legal problems, the most pronounced of which may involve environmental impact. Industrial research project planning should include a comprehensive evaluation of the potential legal aspects of the research.

4.8 RESEARCH QUESTIONNAIRES

Sometimes it is necessary to conduct interviews, surveys and questionnaires as a part of the research effort. These are very useful information gathering techniques. They also foster cooperative working relationships. They encourage direct participation and inputs into project decision making processes. They provide an opportunity for all participants to contribute ideas and inputs to the research. The greater the number of people involved in the interviews, surveys and questionnaires, the more valid the results. The following guidelines are useful for conducting interviews, surveys, and questionnaires to collect data and information for research purposes:

1. Collect and organize background information and supporting documents on the items to be covered by the interview, survey or questionnaire.
2. Outline the items to be covered and list the major questions to be asked.
3. Use a suitable medium of interaction and communication:

telephone, fax, electronic mail, face to face, observation, meeting venue, poster or memo.

4. Tell the respondent the purpose of the interview, survey or questionnaire and indicate how long it will take to complete.
5. Use open-ended questions that stimulate ideas from the respondents.
6. Minimize the use of 'yes' or 'no' type of questions.
7. Encourage expressive statements that indicate the respondent's views.
8. Use the who, what, where, when, why and how approach to elicit specific information.
9. Thank the respondents for their participation.
10. Let the respondents know the outcome of the exercise.

4.9 GROUP DECISION MAKING FOR RESEARCH PLANNING

Many decision situations are complex and poorly understood. No one person has all the information to make all decisions accurately. As a result, crucial decisions are made by a group of people. Some organizations use outside consultants with appropriate expertise to make recommendations for important decisions. Other organizations set up their own internal consulting groups without having to go outside the organization. Decisions can be made via linear responsibility, in which case one person makes the final decision based on inputs from other people. Decisions can also be made through shared responsibility, in which case a group of people share the responsibility for making joint decisions. The major advantages of group decision making are:

- **Ability to share experience, knowledge, and resources**: many heads are better than one. A group will possess greater collective ability to solve a given decision problem.
- **Increased credibility**: decisions made by a group of people often carry more weight in an organization.
- **Improved morale**: personnel morale can be positively influenced because many people have the opportunity to participate in the decision making process.
- **Better rationalization**: the opportunity to observe other people's views can lead to an improvement in an individual's reasoning process.

Some disadvantages of group decision making are:

- It is harder to arrive at a decision. Individuals may have conflicting objectives. For example, while the accounting department representative may want to minimize costs, the labor union representative may push for more use of the work force and protest against the use of automation instead of workers.
- The reluctance of members who oppose the decision to carry it out.
- It hampers future relations between the members of the group, if the tone of the discussion is not controlled.
- Loss of productive employee time.

4.9.1 Brainstorming

Brainstorming is a way of generating many new ideas. In brainstorming, the decision group comes together to discuss alternative ways of solving a problem. The members of the brainstorming group may be from different departments, may have different backgrounds and training, and may not even know one another. The diversity of the participants helps to create a stimulating environment for generating different ideas from different views. The technique encourages free outward expression of new ideas no matter how far-fetched they may appear. No criticism of any new idea is permitted during the brainstorming session. A major concern in brainstorming is that extroverts may take control of the discussions. For this reason an experienced and respected individual should manage the discussions. The group leader establishes the procedure for proposing ideas, keeps the discussions in line with the group's mission, discourages disruptive statements and encourages the participation of all members.

After the group has run out of ideas, open discussions are held to weed out the unsuitable ones. It is expected that even the rejected ideas may stimulate the generation of other ideas which may eventually lead to other more favored ideas. Guidelines for improving brainstorming sessions are presented below:

- Focus on a specific decision problem.
- Keep ideas relevant to the intended decision.
- Be receptive to all new ideas.
- Evaluate the ideas on a relative basis after exhausting new ideas.

- Maintain an atmosphere conducive to cooperative discussions.
- Maintain a record of the ideas generated.

4.9.2 Delphi method

The traditional approach to group decision making is to obtain the opinion of experienced participants through open discussions. An attempt is made to reach a consensus among the participants. However, open group discussions are often biased because of the influence or subtle intimidation of dominant individuals. Even when the threat of a dominant individual is not present, opinions may still be swayed by group pressure. This is called the 'bandwagon effect' of group decision making.

The Delphi method, developed by Gordon and Helmer (1964), attempts to overcome these difficulties by requiring individuals to present their opinions anonymously through an intermediary. The method differs from the other interactive group methods because it eliminates face to face confrontations. It was originally developed for forecasting applications. But it has been modified in various ways for application to different types of decision making. The method can be quite useful for research management decisions. It is particularly effective when decisions must be based on a broad set of factors. The Delphi method is normally implemented as follows:

1. **Problem definition**: a decision problem that is considered significant is identified and clearly described.
2. **Group selection**: an appropriate group of experts or experienced individuals is formed to address the particular decision problem. Both internal and external experts may be involved in the Delphi process. A leading individual is appointed to serve as the administrator of the decision process. The group may operate through the mail or gather together in a room. In either case, all opinions are expressed anonymously on paper. If the group meets in the same room, care should be taken to provide enough room for members not to feel that their responses may be deliberately or accidentally discovered.
3. **Initial opinion poll**: the technique is initiated by describing the problem to be addressed in unambiguous terms. The group members are requested to submit a list of major areas of concern in their specialism areas as they relate to the decision problem.

4. **Questionnaire design and distribution**: questionnaires are pre-pared to address the areas of concern related to the decision problem. The written responses to the questionnaires are collected and organized by the administrator. The administrator aggregates the responses in a statistical format. For example, the average, mode and median of the responses may by computed. This analysis is distributed to the decision group. Each member can then see how his or her responses compare with the anonymous views of the other members.
5. **Iterative balloting**: additional questionnaires based on the previous responses are passed to the members. The members submit their responses again. They may choose to alter or not alter their previous responses.
6. **Silent discussions and consensus**: the iterative balloting may involve anonymous written discussions of why some responses are correct or incorrect. The process is continued until a consensus is reached. A consensus may be declared after five or six iterations of the balloting or when a specified percentage (e.g. 80%) of the group agrees on the questionnaires. If a consensus cannot be declared on a particular point, the whole group may be circulated with a note that it has not produced a consensus.

In addition to its use in technological forecasting, the Delphi method has been widely used in other general decision making. Its major characteristics of anonymity of responses, statistical summary of responses and controlled procedure make it a reliable mechanism for obtaining numeric data from subjective opinion. The major limitations of the Delphi method are:

- Its effectiveness may be limited in cultures where strict hierarchy, seniority and age influence decision making processes.
- Some experts may not readily accept the contribution of non-experts to the group decision making process.
- Since opinions are expressed anonymously, some members may make ludicrous statements. However, if the group composition is carefully reviewed, this problem may be avoided.

4.9.3 Nominal group technique

The nominal group technique is a silent version of brainstorming. It is a method of reaching consensus. Rather than asking people to state their ideas aloud, the team leader asks each member to jot down a minimum number of ideas, for example, five or six. A single list of ideas is then composed on a chalkboard for the whole group to see. The group then discusses the ideas and weeds out some iteratively until a final decision is made. The nominal group technique is easier to control. Unlike brainstorming where members may get into shouting matches, it permits members to present their views silently. In addition, it allows introverted members to contribute to the decision without the pressure of having to speak out too often.

In all these group decision making techniques, an important aspect that can enhance and expedite the decision making process is to require members to review all pertinent data before coming to the group meeting. This will ensure that the decision process is not impeded by trivial preliminary discussions. Some disadvantages of group decision making are:

- Peer pressure in a group situation may influence a member's opinion or discussions.
- In a large group, some members may not participate effectively in the discussions.
- A member's relative reputation in the group may influence how well his or her opinion is rated.
- A member with a dominant personality may overwhelm the other members in the discussions.
- The limited time available to the group may create a time pressure that forces some members to present their opinions without fully evaluating the ramifications of the available data.
- It is often difficult to get all members of a decision group together at the same time.

Despite the noted disadvantages, group decision making definitely has many advantages that may nullify its shortcomings. The advantages as presented earlier will have varying levels of effect from one organization to another. The Triple C principle presented earlier may also be used to improve the success of decision teams. Team work can be enhanced in group decision making by following the guidelines below:

1. Get a willing group of people together.
2. Set an achievable goal for the group.
3. Determine the limitations of the group.
4. Develop a set of guiding rules for the group.

5. Create an atmosphere conducive to group synergism.
6. Identify the questions to be addressed in advance.
7. Plan to address only one topic per meeting.

For major decisions and long-term group activities, arrange for team training which allows the group to learn the decision rules and responsibilities together. The steps for nominal group technique are summarized below:

1. Silently generate ideas, in writing.
2. Record ideas without discussion.
3. Conduct group discussion for clarification of meaning, not argument.
4. Vote to establish the priority or rank of each item.
5. Discuss vote.
6. Cast final vote.

4.9.4 Multivote

Multivoting is a series of votes used to arrive at a group decision. It can be used to assign priorities to a list of items. It can be used at team meetings after a brainstorming session has generated a long list of items. Multivoting helps to reduce such long lists to a few items, usually three to five. The steps for multivoting are:

1. Take a first vote. Each person votes as many times as desired, but only once per item.
2. Circle the items receiving a relatively higher number of votes (i.e. majority vote) than the other items.
3. Take second vote. Each person votes for a number of items equal to one-half the total number of items circled in step 2. Only one vote per item is permitted.
4. Repeat steps 2 and 3 until the list is reduced to three to five items depending on the needs of the group. It is not recommended to multivote down to only one item.
5. Perform further analysis of the items selected in step 4, if needed.

5 | Graduate research management

It is not the load that breaks you down, it is the way you carry it.

The way graduate research is handled will determine how well success is achieved. Graduate research should be managed just like any goal-oriented project. The same techniques of project management are applicable to graduate research projects. However, the major difference is that the major responsibility for the success of the research rests primarily with the graduate student.

5.1 GRADUATE RESEARCH CO-MANAGERS

Find a conducive research advisor.

Consider yourself as the project manager for your graduate research and your advisor as the project director. Relationship with a graduate advisor, particularly at the doctoral level, is usually a life-long commitment. Thus, a good match must be found between the student and the advisor. The research advisor will serve essentially as the project manager for the research effort. Important factors in selecting a research advisor are:

- mutual areas of research interests;
- financial support;
- department's strengths;
- supporting research infrastructure;
- willingness for long-term commitment.

Remember that you are your own best advisor. Based on the findings of your self-assessment, you should be in a position to know exactly what suits you best.

As early as possible, you should identify a faculty member to serve as your research advisor, also called the major professor. This should be approached with caution. It is a sensitive issue that will affect your overall potential for success in the graduate program. Once someone has been officially appointed as your advisor, it may be difficult to change because of complex administrative processes (paperwork and a multitude of signatures), the possibility of impairing research progress or fear of academic reprisal from the 'fired' advisor. The vindictive actions of some professors, given human nature, are one of the realities of graduate research programs that you will need to contend with.

Selection of the research advisor is one of the most crucial decisions a graduate student makes. He or she will become not only the research advisor, but also confidant, mentor, sponsor and advocate throughout the student's graduate program. Many such relationships extend well into the student's own professional career. The research advisor's responsibilities include:

- providing a constructive and supportive research environment;
- helping to justify, rationalize and validate the student's research;
- holding the student to high academic excellence;
- holding the student to high ethical standards;
- serving as the student's advocate within the graduate program;
- facilitating the student's professional development;
- attending to the needs and inquiries of the student promptly;
- reading and commenting on the student's research reports in a timely manner.

Timeliness is the essence of good performance.

Try and gain the commitment of the faculty member that you want to be your research advisor. In some cases, since you are new in the department, the faculty member may need to be convinced of your research match. You may want to furnish him/her with written documentations about your background and research potential. You should arrange 'get-acquainted' meetings with individual faculty members to find out each person's research interests, expertise in a particular area of study, interest in taking on new research advisees, ongoing research projects and potential for funding for a research assistantship.

Remember, the objective is to find a conducive research match regardless of other circumstances. I have seen too many instances where students go with a certain professor simply because of their financial needs rather than their research interests. This is like voting with your nose rather than with your heart. Such students end up frustrated and unproductive in their research efforts. This can lead to serious conflicts with the research advisor. You should do research that you know you will enjoy and be happy doing, because you may end up doing it for the rest of your professional career.

In cases where a professor has advertised for a research assistant to work on a funded research project, you will need to structure your approach just as if you were applying for a regular job. Suggestions and guidelines for preparing for job interviews are presented later in this book.

Even after you find a conducive research advisor, you should not entrust all of your decisions to him or her. Remember that a university catalog exists and you should make extensive use of it to familiarize yourself with the general requirements.

Plan for yourself what you will need to do to accomplish your graduate education objectives.

5.1.1 Establishing a goal

Goals are the basis for planning and action. When evaluated in relation to specific objectives, a goal becomes the road map to success. The characteristics of a good goal are to be:

- clear and specific
- timely
- attainable
- measurable
- prioritized

5.2 EFFECTIVE CONSULTATION WITH THE RESEARCH ADVISOR

Soon after enrollment in a graduate program, you should confer with the director of graduate programs in your department concerning the plan of study. The director will assist you in preparing an initial advisory conference or will direct you to a faculty member for assistance. If a research advisor has already been selected or appointed for you by the department, then you should initiate all your consultations with him or her.

You must recognize that graduate faculty members have diversified responsibilities for teaching, conducting research, writing technical papers, advising students, mentoring graduate students and providing professional service. Understanding these varied responsibilities will help you develop a rapport with the faculty. Because of the nature of their work, the graduate faculty may at first appear withdrawn and indifferent to your needs. But upon closer examination, you will find that what they do is designed to support your graduate study objectives. You should help them as much as they are expected to help you. To develop a sound working relationship with the graduate faculty, particularly your research advisor, you should:

- keep in touch with the faculty without pestering them;
- respect the faculty's busy schedule;
- be prepared for initial self-help before consulting the faculty;
- realize that most minor questions can be answered by the available published guidelines;
- not drop in indiscriminately;
- take the initiative for your own graduate study and research.

5.3 RESEARCH FEASIBILITY

The feasibility of a research project can be ascertained in terms of both technical factors and non-technical factors. A feasibility study is documented with a report showing all the ramifications of the research. It should be noted that not all the factors described below will be pertinent to all engineering and science research projects.

5.3.1 Technical feasibility

Technical feasibility refers to the ability of the research to take advantage of the current state of the technology in pursuing further improvement. The technical capability of the research personnel as well as the capability of the available technology should be considered.

5.3.2 Managerial feasibility

Managerial feasibility involves the capability of the research infrastructure to achieve and sustain intended research objectives. Organizational support, personnel involvement and commitment are key elements required to ascertain managerial feasibility.

5.3.3 Economic feasibility

This involves the feasibility of the proposed project to generate economic benefits. A benefit–cost analysis and a breakeven analysis are important aspects of evaluating the economic feasibility of new industrial projects. The tangible and intangible aspects of a project should be translated into economic terms to facilitate a consistent basis for evaluation.

5.3.4 Financial feasibility

Financial feasibility should be distinguished from economic feasibility. The former involves the capability of the project organization to raise the appropriate funds needed to implement the proposed project. Project financing can be a major obstacle in large multidisciplinary research projects. Access to academic loans, graduate fellowships, tuition waivers and funded research assistantships are important aspects of financial feasibility analysis.

5.3.5 Cultural feasibility

Cultural feasibility deals with the compatibility of the proposed research project with the cultural setup of the research environment. In many industrial projects, planned functions must be integrated with the local cultural practices and beliefs. For example, religious beliefs may dictate what may or may not be included in a research agenda.

5.3.6 Social feasibility

Social feasibility addresses the influences that a proposed project may have on the social system in the project environment. The ambient social structure may be such that certain research results may be controversial and cannot be implemented.

5.3.7 Safety feasibility

Safety feasibility is another important aspect that should be considered in research management, particularly for laboratory-based research. Safety feasibility refers to an analysis of whether the research can be carried out with minimal adverse effects on the environment. Unfortunately, environmental impact assessment is often not adequately addressed in many academic research projects.

5.3.8 Political feasibility

A politically feasible project may be referred to as a 'politically correct project'. Political considerations often dictate the direction for a proposed project. This is particularly true for large projects with national visibility that may have significant government inputs and political implications. For example, political necessity may be a source of support for a project regardless of the project's merits. On the other hand, worthy projects may face opposition because of political factors. Political feasibility analysis requires an evaluation of the compatibility of project goals with the prevailing goals of the political system.

5.3.9 Scope of feasibility analysis

In general terms, the elements of a feasibility analysis for a project should cover the following items:

1. **Need analysis**: this indicates a recognition of a need for the project. The need may affect the organization itself, another organization, the public or the government. A preliminary study is then conducted to confirm and evaluate the need. A proposal as to how the need may be satisfied is then made. Pertinent questions that should be asked include:
   (a) Is the need significant enough to justify the proposed project?
   (b) Will the need still exist by the time the project is completed?
   (c) What are the alternate means of satisfying the need?
   (d) What are the economic, social, environmental, and political impacts of the need?
2. **Engineering analysis**: this is the preliminary analysis carried out to determine what will be required to satisfy the need. The work may be performed by a consultant who is an expert in the

field. The preliminary study often involves system models or prototypes. For technology-oriented projects, an artist's conception and scaled down models may be used to illustrate the general characteristics of a process. A simulation of the proposed system can be carried out to predict the outcome before the actual project starts.

3. **Design**: in this stage, technology capabilities are evaluated and product design is initiated.
4. **Cost estimate**: this involves estimating project cost to an acceptable level of accuracy. Estimates of capital investment, recurring and non-recurring costs should be contained in the cost estimate.
5. **Financial analysis**: this involves an analysis of the cash flow profile of the project. This is an important analysis since it determines whether or not and when funds will be available to continue the project.
6. **Project impacts**: this portion of the feasibility study provides an assessment of the impact of the proposed project. Environmental, social, cultural, political and economic impacts may be some of the factors that will determine how a project is perceived by its patrons.
7. **Conclusions and recommendations**: the feasibility study should end with the overall outcome of the feasibility analysis. Recommendations on what should be done should be included in the report.

5.4 THE TECHNIQUE OF PROJECT MANAGEMENT

Project management is defined as the process of managing, allocating and timing resources to achieve objectives in an efficient and expedient manner. The term project management generally implies the broad conceptual approaches used to manage projects within the constraints of time, cost and quality. These are the same results faced by all graduate students. The quality of a research project may relate to performance requirements or expected technical results.

Organize, prioritize and optimize your research project.

Once you have identified and finalized your research topic, you need to allocate time and resources to the various activities to undertake the research successfully. The conventional tools of project management such as the Gantt chart, responsibility chart and critical path method could be very useful in this regard. The project management process consists of several steps as defined in the following paragraphs.

5.4.1 Problem identification

This is the stage where a need for a proposed project is identified, defined and justified. A project may be concerned with the development of a new theoretical idea, introduction of new product, implementation of new technique, installation of new hardware or any other research and development effort.

5.4.2 Project definition

Project definition is the phase at which the purpose of the project is clarified. A mission statement is the major output of this stage. For example, a prevailing low level of productivity may indicate a need for a new manufacturing technology. In general, the definition should specify how project management may be used to avoid missed deadlines, poor scheduling, inadequate resource allocation, lack of coordination, poor quality and conflicting priorities.

5.4.3 Project planning

A dab of communication can prevent a ton of argument.

A research plan lays out the courses of action for an effective implementation of research ideas. Like any technical undertaking, the research effort must be managed with proven managerial and technical tools. A plan represents the outline of the series of actions needed to accomplish a goal. Project planning determines how to initiate a project and execute its objectives. It may be a simple statement of a project goal or it may be a detailed account of procedures to be followed during the project. Planning can be summarized as:

- objectives
- project definition
- team organization
- performance criteria (time, cost, quality)

5.4.4 Project organizing

Project organization specifies how to integrate the functions of the personnel involved in a project. Organizing is usually done concurrently with project planning. Directing is an important aspect of project organization. Directing involves guiding and supervising the project team and is a crucial aspect of research management. This requires a skillful research director who can interact with subordinates effectively using good communication and motivation techniques. A good project manager can facilitate project success by directing his/her team, through proper task assignments, toward the project goal. Individuals perform better when there are clearly defined expectations; they need to know how their specific tasks contribute to the overall goals of the research.

5.4.5 Resource allocation

Project goals and objectives are accomplished by allocating resources to specific tasks. Resources can consist of money, people, equipment, tools, facilities, information, skills and so on. These are usually in short supply. The individuals needed for a particular task may be committed to other ongoing projects. A crucial piece of laboratory equipment may be under the control of another team. These instances call for a coordinated and cooperative approach to resource allocation.

There is no sense aiming for a goal with no arrow in your bow.

5.4.6 Project scheduling

Scheduling is often the major focus in project management. The main purpose of scheduling is to allocate resources so that the overall project objectives are achieved within a reasonable time span. Project objectives generally conflict. For example, minimization of the project completion time and minimization of the project cost are conflicting objectives. One objective is improved at the expense of the other objective. Scheduling is, thus, a multiple-objective decision making problem.

5.5 PLAN–SCHEDULE–CONTROL CYCLE

Some people plan 'ahead', others plan 'afoot', trusting their goals to the whims of Murphy's law.

The plan–schedule–control (PSC) cycle refers to the process of developing a research plan, following it, evaluating it and taking corrective actions when needed.

In general, scheduling involves the assignment of time periods to specific activities within the work schedule. Resource availability, time limitations, urgency level, required performance level, precedence requirements, work priorities, technical constraints, safety considerations and other factors can complicate the scheduling process. The assignment of a time slot to a task does not necessarily ensure that the task will be performed satisfactorily in accordance with the schedule. Careful control must be developed and maintained throughout the project scheduling. Further details on project scheduling techniques and tools can be found in Badiru and Pulat (1995). Project scheduling involves:

- activity planning;
- resource availability analysis (human, material, money);
- scheduling techniques (CPM, PERT, Gantt charts);
- control.

5.6 PROJECT TRACKING AND REPORTING WITH GANTT CHART

This phase involves checking whether or not project results conform to project plans and performance specifications. Tracking and reporting are prerequisites for project control. A properly organized report of the project status will help identify any deficiencies in the progress of the project and help pinpoint corrective actions. The graphical representation (Gantt chart) presented in Figure 5.1 illustrates both the scheduling and tracking aspects of research activities. The Gantt chart is also referred to as a bar chart or timeline. It shows when each task is to begin and how long it will take to complete.

5.6.1 Project control

Project control requires that appropriate actions be taken to correct unacceptable deviations from expected performance. Control is actuated through measurement, evaluation and corrective action. Measurement is the process of measuring the relationship between planned performance and actual performance with respect to project objectives. The variables to be measured, the measurement

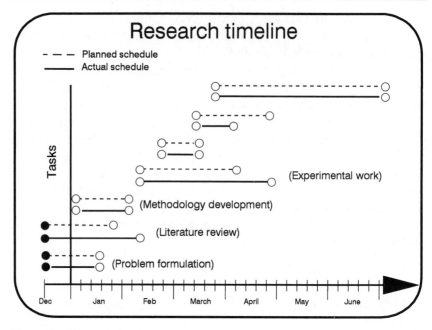

Figure 5.1 Gantt chart of research activities.

scales and the measuring approaches should be clearly specified during the planning stage. Corrective actions may involve rescheduling, reallocation of resources or expediting task performance. Project control involves:

- tracking and reporting;
- measurement and evaluation;
- corrective action (plan revision, rescheduling, updating).

5.6.2 Project termination

This is the last stage of a project. The phase out of a project is as important as its initiation. A project should be terminated expeditiously, and not allowed to drag on after its obvious completion. A terminal activity should be defined for a project during the planning phase. An example of a terminal activity may be the submission of a final report or oral defense of the results. The conclusion of such an activity should be viewed as the completion of the project. Arrangements may be made for follow-up activities that may improve or extend the results. These follow-up or spin-off endeavors should be managed as new projects but with proper input–output relationships within the sequence of projects.

5.7 A PROJECT MANAGEMENT MODEL

An outline of the tasks to be carried out during a research project should be made during the planning stage. A model for such an outline is presented below. This is a generic global model that is suitable for all types of projects (research, industrial, or otherwise). It may be modified and condensed to fit the specific needs of a research project. The model presented is particularly useful for industry-oriented research projects.

Planning

1. Specify project background.
   (a) Define current situation and process:
       (i)   understand the process;
       (ii) identify important variables;
       (iii) quantify variables.
   (b) Identify areas for improvement:
       (i)   list and discuss the areas;
       (ii) study potential strategies for solution.
2. Define unique terminologies relevant to the project:
   (a) discipline-specific terminologies;
   (b) industry-specific terminologies;
   (c) company-specific terminologies;
   (d) project-specific terminologies.
3. Define project goal and objectives:
   (a) write mission statement;
   (b) solicit inputs and ideas from personnel.
4. Establish performance standards:
   (a) schedule;
   (b) performance;
   (c) cost.
5. Conduct formal project feasibility study:
   (a) determine impact on cost;
   (b) determine impact on organization;
   (c) determine project deliverables.
6. Secure management support.

Organizing

1. Identify project management team.
    (a) Specify project organization structure:
        (i)   matrix structure;
        (ii) formal and informal structures;
        (iii) justify structure.
    (b) Specify departments involved and key personnel:
        (i)   purchasing;
        (ii) materials management;
        (iii) engineering, design, manufacturing, etc.
    (c) Define project management responsibilities:
        (i)   select project manager;
        (ii) write project charter;
        (iii) establish project policies and procedures.
2. Implement Triple C model.
    (a) Communication:
        (i)   determine communication interfaces;
        (ii) develop communication matrix.
    (b) Cooperation:
        (i)   outline cooperation requirements.
    (c) Coordination:
        (i)   develop work breakdown structure;
        (ii) assign task responsibilities;
        (iii) develop responsibility chart.

Scheduling and resource allocation

1. Develop master schedule:
    (a) Estimate task duration.
    (b) Identify task precedence requirements:
        (i)   technical precedence;
        (ii) resource-imposed precedence;
        (iii) procedural precedence.
    (c) Use analytical models:
        (i)   CPM;
        (ii) PERT;
        (iii) Gantt chart;
        (iv) optimization models.

Tracking, reporting and control

1. Establish guidelines for tracking, reporting and control.
    (a) Define data requirements:
        (i)   data categories;
        (ii) data characterization;
        (iii) measurement scales.
    (b) Develop data documentation:
        (i)   data update requirements;
        (ii) data quality control;
        (iii) establish data security measures.
2. Categorize control points.
    (a) Schedule audit:
        (i)   activity network and Gantt charts;
        (ii) milestones;
        (iii) delivery schedule.
    (b) Performance audit:
        (i)   employee performance;
        (ii) product quality.
    (c) Cost audit:
        (i)   cost containment measures;
        (ii) percent completion versus budget depletion.
3. Identify implementation process.
    (a) Comparison with targeted schedules.
    (b) Corrective course of action:
        (i)   rescheduling;
        (ii) reallocation of resources.
4. Terminate the project:
    (a) performance review;
    (b) strategy for follow-up projects;
    (c) personnel retention and releases.
5. Document project and submit final report.

5.8 COPING WITH THE STRESS OF RESEARCH

Graduate research can bring on the classical symptoms of stress. Stress must be avoided or controlled to prevent impairment of the research potential. Physical, mental, social, emotional and spiritual alertness can help alleviate the detrimental effects of stress. These can be achieved through a variety of means including exercise, nutrition, reading, positive thinking, writing, goal setting and service to others. Typical effects of stress on the body are:

- tension headache
- muscle pain and spasms
- allergies
- increased blood pressure
- cardiovascular disorders
- frequent heartburn
- constant fatigue
- chronic diarrhea or constipation
- nervousness
- paranoia

Research stress can be handled with simple precautions and/or control actions. Try the following actions to cope with graduate research stress:

- Create opportunities when you can relax your mind and body.
- Engage yourself in enjoyable physical activity as a diversion from stressful activities.
- Prioritize your activities and set reasonable goals.
- Do not expect too much from yourself.
- Avoid agreeing to unachievable deadlines.
- Never play by the deadline (procrastination).
- Eat breakfast and avoid excessive caffeine and sugar intakes.
- Use humor to cope with the demands of research.

5.9 TIME MANAGEMENT

Time management helps you to manage the demands on your time more effectively and efficiently. Through time management, you can accomplish more of those things that are important to you.

Activities that you cannot leave half undone should have priority in your activity schedule (e.g. shaving versus reading newspaper). If you get in a time bind, you can always stop terminable activities and recommence later. To set up a strategy for time and responsibility management, do the following:

1. Evaluate the level of your personal organization.
2. Evaluate the effectiveness and efficiency of your activities. Effectiveness is to do with doing the right things; efficiency refers to doing things right.

Table 5.1 Taxonomy of time wasters

Self-imposed time wasters	*Environmentally imposed time wasters*
1. Over-socializing	1. Delays by others
2. Preoccupation	2. Unproductive environment
3. Attempting to handle too much	3. Unwelcome interruptions
4. Pursuit of perfection	4. Waiting for others' decisions
5. Inability to say no	5. Equipment failure
6. Lack of priorities	6. Others' mistakes
7. Procrastination	7. Lack of policies and procedures
8. Unrealistic time estimates	8. Lack of authority
9. Indecision	9. Lack of feedback
10. Distractions	10. Lack of communication
11. Lack of planning	11. Lack of cooperation
12. Too many mistakes requiring repeats	12. Poor coordination

3. List your long-range personal goals.
4. Develop a comprehensive calendar and personal organizing system.
5. Identify your own time saving approaches.
6. Make time to manage your time. The time invested up front to manage your time effectively will pay off in the attainment of your goals.
7. Identify your major time wasters and determine whether they are imposed by your environment or by your own personal habits, needs, and desires. A list of common time wasters is presented in Table 5.1.

In time management, your objective should not be to command a multitude of activities, but rather to manage your responsibilities so that you do not end up having to do too many things at once. The major concepts and strategies for time management can be summarized as shown below:

- **Activity trap**: focusing on activities themselves rather than the results expected from the activities.
- **80/20 rule**: 80% of your positive results come from the vital 20% of your activities; 20% of total value is associated with the 80% trivial pursuit; 80% of your time is consumed by 20% of value-adding activities. Use this rule to prioritize your activities.
- **Prioritizing**: classifying activities based upon importance as opposed to urgency. Importance should take precedence over urgency.

- **Planning**: personal planning consistently produces better results, and in less time than no planning.
- **Scheduling**: scheduled events are more likely to happen than unscheduled events.
- **Deadlines**: set personal deadlines to inspire action and avoid procrastination.
- **Grouping**: combine similar tasks to facilitate concurrent achievement of results. Yes, you have only two hands, but you can do three things at once. A frequent mistake is to tend to perform activities serially.
- **Daily time management**: manage your time each day by clarifying daily objectives.
- **Clutter avoidance**: avoid clutter so that you can clearly see what needs to be done and when.
- **Cleanup and organize**: always clean up and reorganize before starting any new major task.
- **Smart-up**: work smarter, not necessarily harder.
- **Time budgeting**: allocate your time as if it was the most expensive resource you had.
- **Schedule block**: schedule an entire block of time for one major project. Tune out all trivial activities during that block.
- **Batching**: batch small activities, each of which require a small amount of time, to the end of the day after the major tasks have been accomplished.
- **Deadline or deadblock**: never play by the deadline. Allow time for contingencies prior to the deadline. Last-minute pressure may be challenging, but it is also bankrupting.
- **Formatting**: establish and follow a standard format or rhythm of work that works well with you.
- **Paperless spells**: organize time periods during the day when routine paperwork is not entertained.
- **Easy-does-it**: ease into tough tasks by warming up with simpler activities.
- **Profit hierarchy**: start with the most profitable or important parts of big tasks.
- **Tooling**: have a fixed place for each important tool. Return each to its rightful place promptly after use. This avoids misplacing the tools and forgetting where they are when next they are needed.

5.10 MEETING MANAGEMENT

Meetings are one avenue for information flow for decision making. Effective management of meetings is an important skill needed for achieving goals in a timely manner. Workers often feel that meetings are avenues for wasting time and obstructing productivity. This is because most meetings are poorly organized, improperly managed, called at the wrong time or even unnecessary. In some organizations, meetings are conducted as a matter of routine requirement rather than necessity. Meetings are essential for communication and decision making. Unfortunately, many meetings accomplish nothing and waste everyone's time. A meeting of 30 people wasting only 30 minutes in effect wastes 15 full hours of work time. That much time, in a corporate setting, may amount to thousands of dollars in lost time. It does not make sense to use a one-hour meeting to discuss a task that will take only five minutes to perform. Since meetings (the necessary and the unnecessary) have become an inseparable part of work functions, we must attempt to maximize their output. Some guidelines for running meetings more effectively are presented below:

1. Do pre-meeting homework:
    (a) list topics to be discussed (agenda);
    (b) establish the desired outcome for each topic;
    (c) determine how the outcome will be verified;
    (d) determine who really needs to attend the meeting;
    (e) evaluate the suitability of meeting time and venue;
    (f) categorize meeting topics (e.g. announcements, important, urgent);
    (g) assign time duration to each topic;
    (h) verify that the meeting is really needed;
    (i) consider alternatives to the meeting (e.g. memo, telephone, electronic mail).
2. Circulate the written agenda before the meeting.
3. Start the meeting on time.
4. Review the agenda at the beginning.
5. Get everyone involved; if necessary employ direct questions and eye contact.
6. Keep to the agenda; do not add new items unless absolutely essential.
7. Be a facilitator for meeting discussions.

8. Quickly terminate conflicts that develop from routine discussions.
9. Redirect irrelevant discussions back to the topic of the meeting.
10. Retain leadership and control of the meeting.
11. Recap the accomplishments of each topic before going to the next. Let those who have made commitments (e.g. promise to look into certain issues) know what is expected of them.
12. End the meeting on time.
13. Prepare and distribute minutes. Emphasize the outcome and success of the meeting.

Experimental research guidelines

6

If at first you don't succeed, try again!

Many engineering and science postgraduate researchers use physical laboratory facilities and must interact with technicians and laboratory staff. This often creates tension and difficulties for graduate students. Adversarial relationships can be avoided by following the project management guidelines presented in this book.

6.1 ENHANCING LABORATORY INTERACTIONS

It is very important to interact positively with laboratory technicians and attendants. Graduate students involved in experimentation spend much of their time in the laboratory, often without the direct presence of the supervising professor. With few or no inter-personal skills, the students often get into tense situations with the laboratory staff. This is frequently the result of the mutual disrespect that both groups have for each other. Graduate students may view the lab attendants as low-tech individuals not familiar with the underlying theories of their work. On the other hand, lab attendants may view graduate students as bumbling idiots that do nothing but damage laboratory equipment. Terms such as 'theory with no sense' have been used by technicians to describe graduate students working in the lab.

Laboratory attendants jealously guard their equipment and often restrict access by students. This can seriously impede graduate research that needs laboratory experimentation work. The supervising professors often end up playing both sides of the field, acting as cautious mediators. On one hand, they support the right of the student to have access to lab facilities. On the other hand, they do not want to step on the toes of the lab technicians. The student blames the lab for failed or delayed experiments. The lab blames students for disrupting lab work schedules. The following guidelines should help minimize conflicts in the laboratory:

- Avoid an arrogant approach to the lab attendants.
- Recognize that the lab staff are experts at what they do.
- Respect lab attendants as professionals in their own field.
- Schedule experiments in advance with lab technicians.
- Do not expect the laboratory staff to answer all your questions.
- Recognize that lab staff are limited in what they can do for you.
- Develop a friendly (out-of-lab) rapport with laboratory staff.

The Triple C model presented in an earlier chapter can be used to resolve laboratory conflicts. Such conflicts may include technical conflict, power conflict, schedule conflict, cost conflict, expectation conflict, management conflict, priority conflict, resource conflict and personality conflict. All of these can be alleviated with the Triple C approach through communication, cooperation and co-ordination to facilitate laboratory collaboration.

6.2 THE NEED FOR EXPERIMENTATION

To solve a problem, you sometimes need more information than you have available. You may need to carry out an experimental program to generate the necessary data or information needed for your research. The experiment itself may even be the research. It is important to carry out the experiment efficiently – the objective in developing an experimental model is to make the model as parsimonious as possible. Important questions to address when developing experimentation strategies are:

- Why is the experiment needed?
- Can the information to be generated from the experiment be obtained by other means?
- Are there adequate resources to initiate and carry the experiment through completion?
- Is the required laboratory facility available for the experimental work?

- What are the safety considerations for the experiment?
- Are human and/or animal subjects to be involved in the experiment? If so, what are the institutional safety requirements for such subjects?

6.3 STEPS OF EXPERIMENTAL RESEARCH

1. **Examine the need for the experiment.** Is this retrospective experiment or prospective experiment? A careful examination of the need and nature of the experiment will effectively guide the rest of the experimental work.
2. **Define objectives for the experiment.** Develop and prioritize the things you want the experiment to accomplish. Have a focus! Determine the problem to be solved. Specify the experimental criterion (i.e. dependent variable or response variable). The nature of the criterion helps to determine what statistical tests are applicable.
   (a) Is the response measurable?
   (b) How accurately can it be measured?
   (c) What independent variables are involved?
   (d) Are they measurable?
   (e) Can their levels be manipulated?
   (f) Over what range can they be manipulated?
   (g) Are the levels fixed or random?
3. **Identify the variables of interest.** Most experiments will have many variables, some of which will be of major interest while others will be peripheral. You will need to identify which are the most important in your proposed experiment.
   (a) Is the variable you want to measure really measurable within your experimental scope and capabilities?
   (b) What are the control variables?
   (c) What are the dependent variables?
   (d) What are the independent variables?
   The three ways to handle independent variables are:
   (a) rigidly controlled (extrapolation not recommended);
   (b) manipulated (variable is set at desired levels);
   (c) randomized (levels of variable selected at random).
4. **Identify data categories and classifications.** The type of data involved in your experiment will dictate what you can or cannot do. Different data characteristics may be encountered in an

experimental setting. How do you handle or record each type? What are the really important measurements to make and how to make them? What are the expected ranges of the measurements? What measurement scales will be used?

(a) **The nominal scale** is the lowest level of measurement scale. It classifies items into categories. The categories are mutually exclusive and collectively exhaustive, i.e. they do not overlap and they cover all possible categories of the characteristics being observed. Product type, gender, job classification and color are some examples of measurements on a nominal scale.

(b) **The ordinal scale** is distinguished from a nominal scale by the property of order among the categories. An example is the process of prioritizing tasks for resource allocation. We know that first is above second, but we do not know how far above. Similarly, we know 'better' is preferred to 'good', but we do not know by how much. Data classification into ABC categories is an example of a measurement on an ordinal scale.

(c) **The interval scale** is distinguished from an ordinal scale by having equal intervals between the units of measure. The assignment of data weights from 0 to 10 for a product is an example of a measurement on an interval scale. Even though an item may have a rating of zero, it does not mean that the item has no worthy characteristics at all. Similarly, a score of zero on an examination does not imply that a student knows absolutely nothing about the material covered by the examination. Temperature is a good example of an item that is measured on an interval scale. Even though there is a zero point on the temperature scale, it is an arbitrary relative measure. You cannot touch an item and declare that 'it is zero degrees cold!' Other examples of interval scales are IQ measurements and aptitude ratings.

(d) **The ratio scale** has the same properties of an interval scale, but with a true zero point. For example, time is measured on a ratio scale. Other examples of items measured on a ratio scale are money, volume, length, distance, height, weight and inventory level. Most physical quantities are measured on a ratio scale.

In addition to the measurement scale, data can be classified based on their variability and volatility characteristics. Examples of the relevant classifications are transient data, recurring data, static data and dynamic data.

(a) **Transient data** are defined as a volatile set of data that is encountered once during an expert system consultation and is not needed again. Transient data need not be stored in a permanent database record unless they may be needed for future analysis or uses.

(b) **Recurring data** are encountered frequently enough to necessitate storage on a permanent basis. Recurring data may be further categorized into **static** data and **dynamic** data. Recurring data that are static will retain their original parameters and values each time they are encountered during an experiment. Recurring data that are dynamic have the potential to take on different parameters and values each time they are encountered.

For proper data analysis, data should be recorded in such a way that the structure and format are easy to use and understand. This will enable the analyst to identify possible stratifications and populations that may exist in the data.

5. **Design the experiment.** Determine how observations are taken. Experiments must be properly designed to obtain the maximum amount of information with the minimum expenditure of time and resources. A successfully designed experiment is a series of organized trials which enable you to obtain the most experimental information with the least amount of effort. Pertinent questions for designing experiments are:

(a) What is the minimum number of experiments needed to achieve desired experimental confidence?

(b) What are the protocols for experimental replications?

(c) What are the types of errors to avoid? A type I error is one in which a conclusion is drawn that a variable has an effect on the experimental outcome, when in fact it really does not. A type II error is when the analyst fails to discern a real effect that exists.

Questions addressed by design of experiment are:

(a) How many observations are needed?

(b) Where should the observations be taken?

(c) In what order should the observations be taken?

(d) What variables should be controlled?

(e) What is the randomization procedure?

For example, in an analysis of a family's financial status, if a financial analyst blames the wife's spending habits for the family's poor financial status when, in fact, the poor status is due to the low income of the family, then the analyst has committed a type I error. In an industrial setting, a type I error is often referred to as the producer's risk, while a type II error is referred to as the consumer's risk. In this sense, the former is viewed as the probability of rejecting a good product from a producer while the latter is the probability of a consumer accepting a bad product.

6. **Carry out the experiment.** The minimum number of experiments that must be performed depends on the number of important independent variables in the experiment and how precisely we can measure the results. If there is some error associated with measuring the outcome of an experiment, then it is prudent to repeat the experimental trials. If an experiment is very precise, it is said to be reproducible, and consequently exhibits little error or variation from trial to trial. The less reproducible the experiment and the smaller the error we want to detect, the more data must be collected:

   (a) **reproducibility** (operator variation) is the variation in measurements obtained when several operators use one instrument to measure an identical characteristic on the same part;

   (b) **repeatability** (instrument variation) is the variation in measurements obtained with one instrument when one operator uses it several times to measure an identical characteristic on the same part.

   Care must be exercised in accurately recording experimental results, since repetitions necessitated by careless data recording can be expensive.

7. **Analyze the results.** Use analytical techniques to analyze the results of the experiment. Nowadays, there are numerous software tools available for experimental data analysis. STATGRAPHICS™ and S-PLUS™ are two of the powerful software tools available for desktop computers. These tools permit different views and presentations of the experimental data. Mathematical, graphical and logarithmic representation of experimental data are some of the analytical approaches you should consider for your experiment.

8. **Write a report on the experiment.** The best way to keep up

with experimental results is to document them as they occur. Write your immediate reactions and observations to specific results as they are obtained. These will help you recollect and organize your report later. You should date each item to help with chronological ordering if needed. Follow the report writing guidelines presented in this book to obtain an appealing and convincing report. The crucial contents of your report should be:

(a) **abstract** (a succinct one-page summary indicating your methodology and results);

(b) **introduction** (a general description of the problem area and your problem definition);

(c) **materials and methods**; (description of equipment, materials, and steps of the experiment);

(d) **results** (presentation of the results obtained from the experiment);

(e) **analysis** (description of the analysis performed on the results);

(f) **discussion of findings** (discussion of findings based on the analysis of the results);

(g) **conclusion** (list of important information derived from the experiment);

(h) **references** (list all sources used to supplement the experiment).

9. **Recommend how the results can be used.** Your experimental results are of no use if they cannot be used independently. Describe the features and details of the experimental results as they may be used for decision making purposes. Operational decisions are based on the interpretation of experimental results.

10. **Follow up on the experiment as appropriate.** Your experiment should not be an end in itself. Consider follow-up experiments or activities that may enhance the utility of the experimental results. Investigate how the experimental work can lead to other non-experimental types of research.

6.4 TYPES OF EXPERIMENTATION

Variable manipulation and randomization are the basis for experimentation. The aim is often to infer cause and effect relationships.

Ex post facto research does not involve experimentation. The variables have already acted or interacted to produce the intrinsic results. The researcher only measures, analyzes and interprets what has occurred. Some important concepts to know are:

- **regression**: mathematical equation used for predictive purposes. This is the mathematical model describing the experiment. It expresses the response variable as a function of the independent variables;
- **correlation**: determination of strengths of relationships between variables;
- **double-blind experiment**: the control group does not know that it is the control group and the analyst does not know which is the control group;
- **research hypothesis**: a research hypothesis indicates what the experimenter expects to find in the data.

6.5 DRAWING INFERENCES

Statistical inference is usually divided into two categories: hypothesis testing and estimation. Hypothesis testing involves rejecting or accepting statements about process parameters. Estimation involves estimating the values of process parameters. The discussions on data analysis in the next section deal with estimating parameters such as mean, variance and proportion.

There are two types of hypothesis: the null hypothesis and the alternate hypothesis. The null hypothesis is denoted as H_0 while the alternate hypothesis is denoted as H_1. The null hypothesis is formed primarily to determine if it can be rejected. It represents a probable statement that is viewed as being correct until it is statistically rejected or accepted. The null hypothesis often involves a statement that a parameter is equal to a specified value. The idea of 'no action needed' or 'no difference exists' is often conveyed by the null hypothesis. Hence, the name null hypothesis, where null implies 'no action' or 'no difference' or 'do nothing'.

6.6 HYPOTHESIS TESTING

The implication of rejecting the null hypothesis is that no further action is required either to identify the cause of a difference or to explore the process further. As an example, the statement 'innocent until proven guilty' is a judicial null hypothesis that can be stated as follows:

H_0: The suspect is innocent.
H_1: The suspect is guilty.

Rejecting or accepting the null hypothesis in the above example does not confirm or repudiate the suspect's innocence beyond doubt. Rejection or acceptance is based on available evidence which may not be enough to infer the truth. Similarly, in a statistical evaluation of a process, rejecting or accepting the null hypothesis does not mean that we have arrived at the final conclusion about it. The tentative conclusion may be limited by the quality of the data (evidence) available.

6.6.1 One-tailed versus two-tailed hypothesis testing

A test of hypothesis may be one-tailed or two-tailed depending on the direction of the statement contained in the alternate hypothesis. An example of a two-tailed test is:

H_0: process average $= 200$
H_1: process average $\neq 200$

In the above example, H_0 will be rejected if the process average is above or below 200. The one-tailed tests for the example are explained below:

H_0: process average $= 200$
H_1: process average > 200

In this case, H_0 will be rejected only if the process average is above 200.

H_0: process average $= 200$
H_1: process average < 200

In this case, H_0 will be rejected only if the process average is below 200.

Hypothesis testing can be carried out to evaluate a single process based on a sample from the process. This is referred to as a one-sample study. If the study is done to compare two processes based on samples drawn from the processes, then the study is referred to as a two-sample study.

The significance level in hypothesis testing is expressed in terms of α. For example, an alpha value of 0.001 ($\alpha = 0.001$) implies taking one chance in 1000 of rejecting H_0 when it is true, i.e. probability of rejecting H_0 when it is true (type I error risk level). The beta value β refers to the probability of accepting H_0 when it is false (type II error). The power of the test $(1 - \beta)$ refers to the probability of rejecting H_0 when it is false. In practical terms, type II error is more damaging because it involves accepting a false statement. The power of the test may be increased by:

- increasing sample size
- increasing the alpha level

Increasing α means that the significance level $(1 - \alpha)$ will be decreased. Thus, the test will be more effective in detecting differences. The tighter the tolerance, the more difficult it is for the test to be effective or powerful. The operating characteristic curve is a plot of beta versus the population mean (β versus μ). The power curve is a plot of power versus population mean $(1 - \beta$ versus μ). Complete randomization averages out time or factor dependent effects on an experiment. Table 6.1 presents a summary of the experimental process.

Table 6.1 Summary of experimental process

1. Experiment:
    (a) Statement of problem
    (b) Choice of response variable
    (c) Selection of factors to be varied
    (d) Choice of factor levels
        (i) Quantitative versus qualitative
        (ii) Fixed versus random

2. Design
    (a) Number of observations to be taken
    (b) Order of experimentation
    (c) Method of randomization to be used
    (d) Mathematical model to describe the experiment
    (e) Specification of hypothesis to be tested

3. Analysis
    (a) Data collection and processing
    (b) Computation of test statistics
    (c) Interpretation of results

6.7 VERIFICATION AND VALIDATION

Research must be verified and validated before being implemented in a functional system. Without proper verification and validation, results can be disappointing. An important reason for performing careful verification is that when a system malfunctions, the source of the problem may not be as obvious as it is with conventional programs. Thus, the problem may go undetected until serious harm has been done.

6.7.1 What is verification?

Verification is the determination of whether or not the system is functioning as intended. This may involve program debugging, error analysis, input acceptance check, output verification, reasonableness of operation, run time control and result documentation.

6.7.2 What is validation?

Validation is a diagnosis of how closely a system's solution matches the solution expected. If the system is valid, then the decisions, conclusions or recommendations it offers can form the basis for setting actual operating conditions. The validation should be done by using different problem scenarios to simulate actual system's operations.

It may sometimes be impossible to validate a system for all the anticipated problems because the data for such problems may not yet be available. In such a case, the closest possible representation of the expected scenarios should be utilized for the validation.

What to validate

The crucial characteristic components of a system should be identified and used as the basis for validation. For example, if your research involves the development of an expert system, then the knowledge base component of the system is the area which will require the most thorough evaluation since it contains the problem-solving strategies of the expert system.

How much to validate

The methodology that is used to determine how much validation to perform depends on the number of representative cases which are available for evaluation. For example, evaluating a medical knowledge base which diagnoses rare diseases will be much more difficult than evaluating one which addresses a common problem. Depending on the availability of representative cases, special techniques may need to be used to perform validation (i.e. sensitivity analysis, what if analysis).

The degree of validation to be performed on a system depends on the degree of significance attached to it. This is assessed from the context in which the system will be used. In some cases, a system may be viewed as a complete replacement for a human operator. This implies total dependence on the system and a consequent need for greater validation. In other cases, a system may be developed as a complementary tool in problem solving. This type of use does not require very rigorous validation.

When to validate

The appropriate time to perform validation is a key decision in any process. Since errors can occur anywhere in the development process from data collection to methodology development, validating a system in stages is very important; this makes it easier to catch errors before they become compounded. For very small systems, validation can be performed in one single stage at the completion of the system.

6.7.3 Verification and validation process

In a very large system, validation should occur at each stage of the development cycle. Small and medium systems may be validated at only a few selected intervals. Recommended stages for validation are:

1. **Conceptualization stage** with overall goal definition: state what the measures of the program's success will be and how failure or success will be evaluated.
2. **First version prototype** showing feasibility: demonstrate feasibility of the system and perform preliminary evaluation with a few special test cases.
3. **System refinement**: evaluate with informal test cases and get feedback from experts and possible end users.
4. **Evaluation of performance**: perform formalized evaluation using randomly selected data inputs.

5. **Evaluation of acceptability to users**: evaluate the system in its intended users' surroundings. Verify that the system has good human factors concepts (i.e. input–output devices and ease of use).
6. **Evaluation of functionality** for extended period in prototype environment: field test and verify the system. Observe performance of the system and reactions of the users.
7. **Pre-implementation evaluation**: evaluate the overall system prior to deployment in an operating environment.

Factors involved in validation

Several major factors should be examined carefully in the verification and validation stages of a system. The objective is to verify and validate that for any correct input to the system, a correct output can be obtained. Factors of interest in system validation include:

- **Completeness** refers to the thoroughness of the system and checks if the system can address all desired problems within its problem domain.
- **Efficiency** checks how well the system makes uses of the available knowledge, data, hardware, software, and time in solving problems within its specified domain.
- **Validity** is the correctness of the system outputs. Validity may be viewed as the ability of the system to provide accurate results for relevant data inputs.
- **Maintainability** is how well the integrity of the system can be preserved even when operating conditions change.
- **Consistency** requires that the system provide similar results to similar problem scenarios.
- **Precision** refers to the level of certainty or reliability associated with the consultations provided by the system. Precision is often application dependent. For example, precision in a medical diagnosis may be more important than precision in other diagnostic domains. Compliance with any prevailing rules and regulations is an important component of the precision of a technical system.
- **Soundness** refers to the quality of the scientific and technical basis for the methodology of the system.
- **Usability** involves an evaluation of how the system might meet users' needs. Questions to be asked include whether the system

is usable by the end user. Are questions worded in an easily understood format? Is help available? Is the system able to explain its reasoning process to the user? Is the system compatible with the delivery environment?

- **Justification** is a key factor of validation. A system should be justified in terms of cost requirements, operating characteristics, maintainability and responsiveness to user requests.
- **Reliability**: under reliability evaluation, the system is expected to perform satisfactorily whenever it is used. It should not exhibit erratic performance and results. Several test runs are typically needed to ascertain the reliability of a system.
- **Accommodating**: the system has to be forgiving of minor data entry errors by the user. Appropriate prompts should be incorporated into the user interface to inform users of incorrect data inputs and allow correction of inputs.
- **Clarity** refers to how well the system presents its prompts to avoid ambiguities in the input/output processes. If the system possesses a high level of clarity, there will be assurance that it will be used as intended by the users.
- The **quality** of a system is the subjective perception of users of the system. Quality is often defined as a measure of the user's satisfaction. It refers to the comprehensive combination of the characteristics of a system that determines the system's ability to satisfy specific needs.

How to evaluate the system

To correctly validate a system on the basis of empirical analysis, the correct results for test cases must be known and accessible. With known results, an absolute measure of the effectiveness of the system may be estimated as the proportion of correct to incorrect results produced by the system. If standard results are not available, then a relative evaluation of the system may be performed on the basis of the performance of other systems designed to perform similar functions. These are some guidelines on how to perform the evaluation process:

- Set realistic standards for the performance of the system.
- Define the minimum acceptable standard required for the system to be considered successful.
- Use performance standards that are comparable to those used in evaluating comparable systems.

- Use controlled experiments where the evaluators are not biased by the sources of the results being evaluated.
- Distinguish between 'false positive' and 'true positive' results produced by the expert system. In a false positive result, the system would diagnose as 'true' what is not really 'true'. In true positive results, the system will diagnose as 'true' only what is really 'true'.
- In cases of incorrect results, identify which correct solutions are closest to being reached. This will be very valuable in performing a refinement of the system later on.

6.7.4 Sensitivity analysis

To improve the precision of a system, the researcher can carry out sensitivity analysis. This establishes the variability in the conclusions of the system as a function of the variability of the data, i.e. it identifies the differences in results that are caused by different levels of change in the input to the system. If minor changes in the inputs lead to large differences in the result then the system is said to be very sensitive to changes in inputs. One effective method of using sensitivity analysis to improve precision is to display as a histogram output values against possible answers for one given input. Sensitive points will be displayed as significant changes in the histogram. These visual identifications help identify potential trouble spots in the system.

7 | Data analysis and presentation

Data, data everywhere; not a byte to use!

Data analysis is the various mathematical and graphical operations that can be performed on experimental data to elicit the inherent information contained in the data. Analysis involves the procedure for data collection, organization, reduction and computational processing. The manner in which data is analyzed and presented can significantly affect the way the information is perceived by the decision maker.

7.1 DATA ORGANIZATION

Misrepresented or misinterpreted data can lead to erroneous conclusions and decisions. Different techniques are available for analyzing the different types of data. We will use the sample data in Table 7.1 to illustrate some simple data analysis techniques.

Table 7.1 Experimental data comparing four processes

Process	Yield from run 1	Yield from run 2	Yield from run 3	Yield from run 4	Row total
A	3 000	3 200	3 400	2 800	12 400
B	1 200	1 900	2 500	2 400	8 000
C	4 500	3 400	4 600	4 200	16 700
D	2 000	2 500	3 200	2 600	10 300
Column total	10 700	11 000	13 700	12 000	47 400

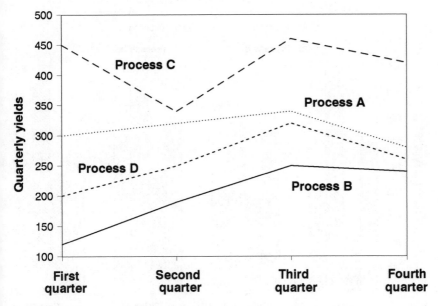

Figure 7.1 Line graph of process yields.

7.1.1 Raw data

Raw data consist of the actual observations recorded about a process attribute. Raw data should be organized into a format suitable for visual review and computational analysis. The data in Table 7.1 represent experimental yields from four processes. Figure 7.1 presents the raw data as a line graph over time. The same information is presented as a multiple bar chart in Figure 7.2.

7.2 DATA MEASURES AND ATTRIBUTES

Different measures and attributes are used to convey the characteristics of data. The most common measures are presented below.

7.2.1 Data total

Total or sum is a measure that indicates the overall effect of a particular variable. If X_1, X_2, X_3, ..., X_n represent a set of n observations (e.g. yields), then the total is computed as:

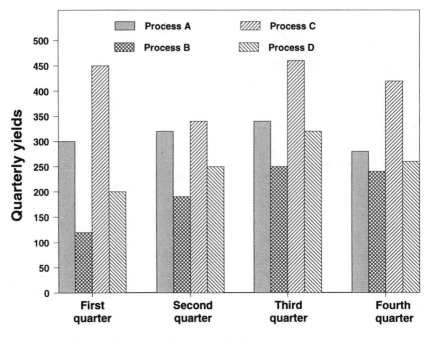

Figure 7.2 Multiple bar chart of process yields.

$$T = \sum_{i=1}^{n} X_i$$

For the sample data, the total yield for each process is shown in the last column. The totals indicate that process C produced the largest total yield over the four runs while process B produced the lowest total yield. The last row of the table shows the total yield for each run. The totals reveal that the largest yield occurred in the third run. The first run brought in the lowest total yield. The grand total yield for the four processes over the four runs is shown as 47 400 in the last cell in the table. Figure 7.3 presents the total yields by run as stacked bar charts. Each segment in a stack of bars represents the yield from a particular process. The total yields for the four processes over the four runs are shown in a pie chart in Figure 7.4. The percentage of the overall yield contributed by each process is also shown in the pie chart.

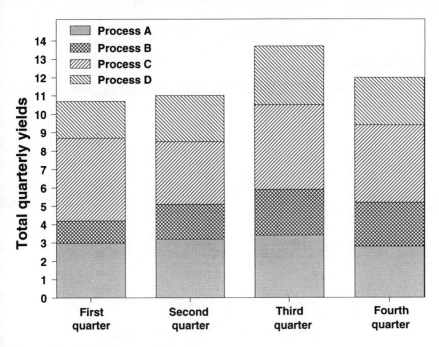

Figure 7.3 Stacked bar graph of run totals.

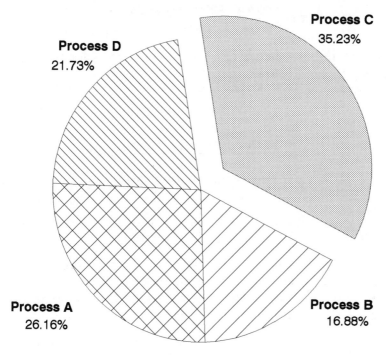

Figure 7.4 Pie chart of total yield per process.

7.2.2 Average

Average is one of the most used measures in data analysis. Given n observations (e.g. yield), $X_1, X_2, X_3, \ldots, X_n$, the average is computed as:

$$\bar{X} = \frac{\sum_{i=1}^{n} X_i}{n}$$

$$= \frac{T}{n}$$

For the data in Table 7.1, the average run yields for the four processes are:

$$\bar{X}_A = \frac{(3000 + 3200 + 3400 + 2800)}{4} = 3100$$

$$\bar{X}_B = \frac{(1200 + 1900 + 2500 + 2400)}{4} = 2000$$

$$\bar{X}_C = \frac{(4500 + 3400 + 4600 + 4200)}{4} = 4175$$

$$\bar{X}_D = \frac{(2000 + 2500 + 3200 + 2600)}{4} = 2575$$

Similarly, the expected average yield per process for the four runs are as presented below:

$$\bar{X}_1 = \frac{(3000 + 1200 + 4500 + 2000)}{4} = 2675$$

$$\bar{X}_2 = \frac{(3200 + 1900 + 3400 + 2500)}{4} = 2750$$

$$\bar{X}_3 = \frac{(3400 + 2500 + 4600 + 3200)}{4} = 3425$$

$$\bar{X}_4 = \frac{(2800 + 2400 + 4200 + 2600)}{4} = 3000$$

The above values are shown in a bar chart in Figure 7.5. The average yield from any of the four processes at any given run is calculated as the sum of all the observations divided by the number of observations. That is,

$$\bar{X} = \frac{\sum_{i=1}^{N} \sum_{j=1}^{M} X_{ij}}{K}$$

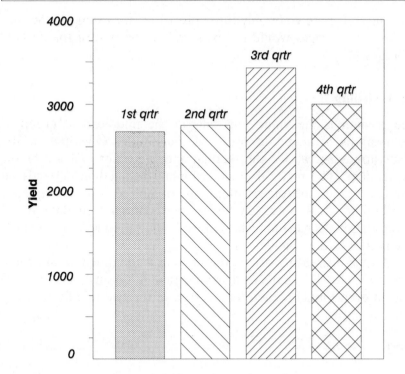

Figure 7.5 Average expected yield per process for each run.

where
N = number of processes
M = number of runs
K = total number of observations $(K = NM)$

For the sample data, overall average per process per run is:

$$\bar{X} = \frac{47\,400}{16}$$

$$= 2962.50$$

As a cross-check, the sum of the run averages should be equal to the sum of the process yield averages, which is equal to the grand total divided by four, i.e.

$$(2675 + 2750 + 3425 + 3000) = (3100 + 2000 + 4175 + 2575)$$

$$= 11\,800$$

The cross-check procedure above works because we have a balanced table of observations. That is, we have four processes and four runs. If there were only three processes, for example, the sum of the run averages would not be equal to the sum of the process averages.

7.2.3 Median

The median is the value that falls in the middle of a group of observations arranged in order of magnitude. One half of the observations are above the median and the other half are below the median. The method of determining the median depends on whether or not the observations are organized into a frequency distribution. It is necessary to arrange unorganized data in an increasing or decreasing order before finding the median. Given K observations (e.g. yields), X_1, X_2, X_3, ..., X_K, arranged in increasing or decreasing order, the median is identified as the value in position $(K+1)/2$ in the data arrangement if K is an odd number. If K is an even number, then the average of the two middle values is

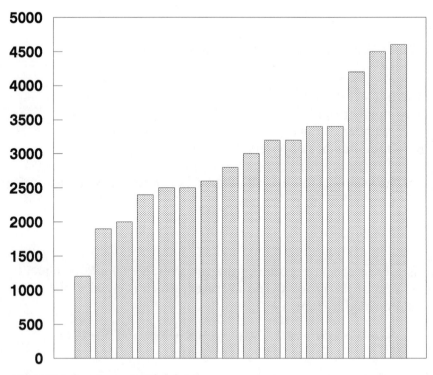

Figure 7.6 Bar chart of ordered data.

considered to be the median. If the sample data are arranged in increasing order, we would get the following:

1200, 1900, 2000, 2400, 2500, 2500, 2600, 2800, 3000, 3200, 3200, 3400, 3400, 4200, 4500, 4600

The median is then calculated as $(2800 + 3000)/2 = 2900$. Thus, half of the recorded yields are expected to be above 2900 while half are expected to be below. Figure 7.6 presents a bar chart of the yield data arranged in increasing order. The median is anywhere between the eighth and ninth values in the ordered data above.

7.2.4 Quartiles and percentiles

The median discussed above is a position measure because its value is based on its position in a set of observations. Other measures of position are quartiles and percentiles. There are three quartiles which divide a set of data into four equal categories. The first quartile, denoted Q1, is the value below which one-fourth (one run) of all the observations in the data set fall. The second quartile, denoted Q2, is the value below which two-fourths or one-half of all the observations in the data set fall. The third quartile, denoted Q3, is the value below which three-fourths of the observations fall. The second quartile is identical to the median. It is technically incorrect to talk of the fourth quartile because that will imply that there is a point within the data set below which all the data points fall: a contradiction! A data point cannot lie within the range of the observations and at the same time exceed all the observations, including itself.

The concept of percentiles is similar to the concept of quartiles except that reference is made to percentage points. There are 99 percentiles that divide a set of observations into 100 equal parts. The X percentile is the value below which X% of the data fall. Thus, the 99 percentile refers to the point below which 99% of the observations fall. The three quartiles discussed previously are regarded as the 25th, 50th and 75th percentiles. As was explained above, it would be technically incorrect to talk of the 100 percentile. In performance ratings, such as product quality evaluation, the higher the percentile of a product, the better the product. In many cases, recorded data are classified into categories that are not indexed to numerical measures. In such cases, other measures of central tendency or position will be needed. An example of such a measure is the mode.

7.2.5 The mode

The mode is defined as the value that has the highest frequency in a set of observations. When the recorded observations can only be classified into categories, the mode can be particularly helpful in describing the data. Given a set of K observations (e.g. yields), X_1, X_2, X_3, ..., X_K, the mode is identified as that value that occurs more than any other value in the set. Sometimes the mode is not unique in a set of observations. For example, in our sample data, 2500, 3200 and 3400 all have the same number of occurrences (two each). Thus, each of them is a mode of the set of yield observations. If there is a unique mode in a set of observations, then the data are said to be unimodal. The mode is very useful in expressing the central tendency of a process having qualitative characteristics such as color.

7.2.6 Range

The range is determined by the two extreme values in a set of observations. Given K observations (e.g. yields), X_1, X_2, X_3, ..., X_K, the range of the observations is simply the difference between the lowest and the highest observations. This measure is useful when one wants to know the extent of extreme variations in a factor such as process yield. The range of the yields in the sample data is $(4600 - 1200) = 3400$. Because of its dependency on only two values, the range tends to increase as the sample size increases. Furthermore, it does not provide a measurement of the variability of the observations relative to the center of the distribution. This is why the standard deviation is used as a more reliable measure of dispersion than the range.

The variability of a distribution is generally expressed in terms of the deviation of each observed value from the sample average. If the deviations are small, the set of data is said to have low variability. The deviations provide information about the degree of dispersion in a set of observations. Unfortunately, a general formula to evaluate the variability of data cannot be based on the deviations. This is because some of the deviations are negative while some are positive and the sum of all the deviations is equal to zero. One possible solution to this problem is to compute the average deviation.

7.2.7 Average deviation

The average deviation is the average of the absolute values of the deviations from the sample average. Given K observations, X_1, $X_2, X_3, \ldots, X_K$, the average deviation of the data is computed as:

$$\bar{D} = \frac{\sum\limits_{i=1}^{K} |X_i - \bar{X}|}{K}$$

Table 7.2 shows how the average deviation is computed for the sample data. One of the most serious disadvantages of the average of deviation measure is that the procedure ignores the sign associated with each deviation. Despite this disadvantage, its simplicity and ease of computation makes it especially useful to users with limited knowledge of statistical methods. In addition, a knowledge of the average deviation helps in understanding the standard deviation, which is a very important measure of dispersion.

Table 7.2 Table of average deviation, standard deviation and variance

Observation Number (i)	Recorded value X_i	Deviation from average $X_i - \bar{X}$	Absolute deviation $|X_i - \bar{X}|$	Square of Deviation $(X_i - \bar{X})^2$
1	3000	37.5	37.5	1406.25
2	1200	−1762.5	1762.5	3 106 406.30
3	4500	1537.5	1537.5	2 363 906.30
4	2000	−962.5	962.5	926 406.25
5	3200	237.5	237.5	56 406.25
6	1900	−1062.5	1062.5	1 128 906.30
7	3400	437.5	437.5	191 406.25
8	2500	−462.5	462.5	213 906.25
9	3400	437.5	437.5	191 406.25
10	2500	−462.5	462.5	213 906.25
11	4600	1637.5	1637.5	2 681 406.30
12	3200	237.5	237.5	56 406.25
13	2800	−162.5	162.5	26 406.25
14	2400	−562.5	562.5	316 406.25
15	4200	1237.5	1237.5	1 531 406.30
16	2600	−362.5	362.5	131 406.25
Total	47 400.0	0.0	11 600.0	13 137 500.25
Average	2962.5	0.0	725.0	821 093.77
Square root				906.14

7.2.8 Sample variance

Sample variance is the average of the squared deviations computed from a set of observations. If the variance of a set of observations is large, the data are said to have a large variability. For example, a large variability in process performance may indicate a lack of consistency or improper methods in the process. Given K observations, X_1, X_2, X_3, . . . , X_K, the sample variance of the data is computed as:

$$s^2 = \frac{\sum_{i=1}^{K} (X_i - \bar{X})^2}{K - 1}$$

The variance can also be computed by the following alternate formulae:

$$s^2 = \frac{\sum_{i=1}^{K} X_i^2 - \left(\frac{1}{K}\right)\left[\sum_{i=1}^{K} X_i\right]^2}{K - 1}$$

$$s^2 = \frac{\sum_{i=1}^{K} X_i^2 - K(\bar{X})^2}{K - 1}$$

Using the first formula, the sample variance of the data in Table 7.2 is calculated as:

$$s^2 = \frac{13\,137\,500.25}{16 - 1}$$

$$= 875\,833.33$$

Note that the average calculated in the last column of Table 7.2 is obtained by dividing the total for that column by 16 instead of 16 $- 1 = 15$. Thus, that average is not the correct value of the sample variance. However, as the number of observations becomes very large, the average as computed in Table 7.2 will become a close estimate for the correct sample variance. Statisticians generally distinguish between the two values by referring to the average calculated in Table 7.2 as the population variance when K is very large and referring to the average calculated by the formulae above as the sample variance particularly when K is small. Thus, for our example, the population variance is given by:

$$\sigma^2 = \frac{\sum\limits_{i=1}^{K}(X_i - \bar{X})^2}{K}$$

$$= \frac{13\,137\,500.25}{16}$$

$$= 821\,093.77$$

while the sample variance, as shown previously, is given by:

$$s^2 = \frac{\sum\limits_{i=1}^{K}(X_i - \bar{X})^2}{K - 1}$$

$$= \frac{13\,137\,500.25}{(16 - 1)}$$

$$= 875\,833.33$$

The tabulation of the raw data and the computations shown in Table 7.2 can be quite laborious. For this reason, software tools such as spreadsheet programs are recommended for experimental data analysis.

7.2.9 Standard deviation

The sample standard deviation of a set of observations is the positive square root of the sample variance. The use of variance as a measure of variability has some drawbacks. For example, the knowledge of the variance is helpful only when two or more sets of observations are compared. As a result of the squaring operation, the variance is expressed in square units rather than the original units of the raw data. To get a reliable feel for the variability in the data, it is necessary to restore the original units by performing the square root operation on the variance. This is why standard deviation is a widely recognized measure of variability. Given K observations, $X_1, X_2, X_3, \ldots, X_K$, the sample standard deviation of the data is computed as:

$$s = \sqrt{\frac{\sum\limits_{i=1}^{K}(X_i - \bar{X})^2}{K - 1}}$$

As in the case of the sample variance, the sample standard deviation can also be computed by the following alternate formulae:

$$s = \sqrt{\frac{\sum\limits_{i=1}^{K} X_i^2 - \left(\frac{1}{K}\right)\left[\sum\limits_{i=1}^{K} X_i\right]^2}{K - 1}}$$

$$s = \sqrt{\frac{\sum\limits_{i=1}^{K} X_i^2 - K(\bar{X})^2}{K - 1}}$$

Using the first formula, the sample standard deviation of the data in Table 7.2 is calculated as:

$$s = \sqrt{\frac{13\,137\,500.25}{16 - 1}}$$

$$= \sqrt{875\,833.33}$$

$$= 935.8597$$

Thus, we can say that the variability in the expected yield per process per run is 935 859.70. As was previously explained for the sample variance, the population sample standard deviation is given by:

$$\sigma = \sqrt{\frac{\sum\limits_{i=1}^{K} (X_i - \bar{X})^2}{K}}$$

$$= \sqrt{\frac{13\,137\,500.25}{16}}$$

$$= \sqrt{821\,093.77}$$

$$= 906.1423$$

while the sample standard deviation is given by:

$$s = \sqrt{\frac{\sum\limits_{i=1}^{K} (X_i - \bar{X})^2}{K - 1}}$$

$$= \sqrt{\frac{13\,137\,500.25}{(16 - 1)}}$$

$$= \sqrt{875\,833.33}$$

$$= 935.8597$$

7.3 TYPES OF STATISTICS

7.3.1 Descriptive statistics

Descriptive statistics are analyses that are performed in order to describe the nature of the results obtained from an experiment. The analyses presented previously fall under the category of descriptive statistics because they are concerned with summary calculations and graphical display of observations.

7.3.2 Inferential statistics

Inferential statistics are derived from the process of drawing inferences about a process on the basis of limited observations. The techniques presented below fall under the category of inferential statistics. Inferential statistics are of more interest to many experimental analysts because they are more dynamic and provide generalizations about a population by investigating only a portion of it. The portion of the population investigated is referred to as a sample. As an example, the expected duration of a proposed task can be inferred from several previous observations of the durations of identical tasks. Other sets of tools used in statistical analysis are deductive statistics and inductive statistics.

7.3.3 Deductive statistics

Deductive statistic involve assigning properties to a specific item in a set based on the properties of a general class covering the set. For example, if it is known that 90% of processes in a given organization fail, then deduction can be used to assign a probability of 90% to the event that a specific process in the organization will fail.

7.3.4 Inductive statistics

Inductive statistic involve drawing general conclusions from specific facts, i.e. inferences about populations are drawn from samples. For example, if 95% of a sample of 100 people surveyed in a 5000-person organization favor a particular process, then induction can be used to conclude that 95% of the personnel in the organization favor the process.

7.4 SAMPLES AND SAMPLING

A sample space of an experiment is the set of all possible distinct outcomes of the experiment. An experiment is some process that generates distinct sets of observations. The simplest and most common example is the experiment of tossing a coin to observe whether heads or tails will show up. An outcome is a distinct observation resulting from a single trial of an experiment. In the experiment of tossing a coin, 'heads' and 'tails' are the two possible outcomes. Thus, the sample space consists of only two items.

There are numerous examples of statistical experiments in science and engineering research. An experiment may involve simply checking to see whether it rains or not on a given day. Another experiment may involve counting how many tasks fall behind schedule during a process. Another example of an experiment may involve recording how long it takes to perform a given activity in each of several trials. The outcome of any experiment is frequently referred to as a random event because the outcomes of the experiment occur in a random fashion. We cannot predict with certainty what the outcome of a particular trial of the experiment will be.

7.4.1 Sample

A sample is a subset of a population that is selected for observation and statistical analysis. Inferences are drawn about the population based on the results of the analysis of the sample. The reasons for using sampling rather than complete population enumeration are:

1. It is more economical to work with a sample.
2. There is a time advantage to using a sample.
3. Populations are typically too large to work with.
4. A sample is more accessible than the whole population.
5. In some cases, the sample may need to be destroyed during the experiment.

There are three primary types of samples. They differ in the manner in which their elementary units are chosen.

Convenience sample

A convenience sample refers to a sample that is selected on the basis of how convenient certain elements of the population are for observation. Convenience may be needed because of time pressures, process accessibility or other considerations.

Judgment sample

A judgment sample is one that is obtained based on the discretion of someone familiar with the relevant characteristics of the population. This is usually based on the heuristics of an experienced individual. It may also be based on a group consensus.

Random sample

A random sample is a sample where the elements are chosen at random. This is the most important type of sample for statistical analysis. In random sampling, all the items in the population have an equal chance of being selected for inclusion in the sample.

Since a sample is a collection of observations representing only a portion of the population, the way in which the sample is chosen can significantly affect the adequacy and reliability of the sample. Even after the sample has been chosen, the manner in which specific observations are obtained may still affect the validity of the results. The possible bias and errors in the sampling process are discussed below.

Sampling and non-sampling errors

A sampling error is a difference between a sample mean and a population mean that arises from the particular sample elements that are selected for observation. A non-sampling error refers to an error arising from the manner in which the observation is made.

Sampling bias

A sampling bias is the tendency to favor the selection of certain sample elements with specific characteristics. For example, a sampling bias may occur if a sample of the personnel is taken only from the engineering department in a survey addressing the implementation of high technology processes.

Stratified sampling

Stratified sampling involves dividing the population into classes, or groups, called strata. The items contained in each stratum are expected to be homogeneous with respect to the characteristics to be studied. A random subsample is taken from each stratum. The subsamples from all the strata are then combined to form the desired overall sample. Stratified sampling is typically used for a heterogeneous population such as employees in an organization. Through stratification, groups of employees are set up so that the individuals within each stratum are mostly homogeneous and the strata are different from one another. As another example, a survey of managers on some important issue of worker involvement may be conducted by forming strata on the basis of the types of processes they are involved with. There may be one stratum for technical processes, one for construction processes and one for manufacturing processes.

A proportionate stratified sampling results if the units in the sample are allocated among the strata in proportion to the relative number of units in each stratum in the population, i.e. an equal sampling ratio is assigned to all strata in a proportionate stratified sampling. In disproportionate stratified sampling, the sampling ratio for each stratum is inversely related to the level of homogeneity of the units in the stratum. The more homogeneous the stratum, the smaller its proportion included in the overall sample. The rationale for using disproportionate stratified sampling is that when the units in a stratum are more homogeneous, a smaller subsample is needed to ensure good representation. The smaller subsample helps to reduce sampling cost.

Cluster sampling

Cluster sampling involves the selection of random clusters, or groups, from the population. The desired overall sample is made up of the units in each cluster. Cluster sampling is different from stratified sampling in that differences between clusters are usually small. In addition, the units within each cluster are generally more heterogeneous. Each cluster, also known as the primary sampling unit, is expected to be a scaled down model that gives a good representation of the characteristics of the population.

All the units in each cluster may be included in the overall sample or a subsample of the units in each cluster may be used. If all the units of the selected clusters are included in the overall sample, the procedure is referred to as a single-stage sampling. If a subsample is taken at random from each selected cluster and all units of each subsample are included in the overall sample, then the sampling procedure is called a two-stage sampling. If the sampling procedure involves more than two stages of subsampling, then the procedure is referred to as a multistage sampling. Cluster sampling is typically less expensive to implement than stratified sampling. For example, the cost of taking a random sample of 2000 managers from different industry types may be reduced by first selecting a sample, or cluster, of 25 industries and then selecting 80 managers from each of the 25 industries. This represents a two-stage sampling that will be considerably cheaper than trying to survey 2000 individuals in several companies in a single-stage procedure.

7.5 FREQUENCY DISTRIBUTION

Once a sample has been drawn and observations of all the items in the sample are recorded, the task of data collection is completed. The next task involves organizing the raw data into a meaningful format. In addition to the various methods discussed earlier, frequency distribution is another tool for organizing data. Frequency distribution involves the arrangement of observations into classes to show the frequency of occurrences in each class. An appropriate class interval must be selected for the construction of the frequency distribution. The guidelines for selecting the class interval are:

1. The number of classes should not be too small or too large that the true nature of the underlying distribution cannot be identified. Generally, the number of classes should be between 6 and 20.
2. The interval length of each class should be the same. The interval length should be selected such that every observation falls within some class.
3. The difference between midpoints of adjacent classes should be constant and equal to the length of each interval.

Suppose an experiment yields the following 20 observations:

3000, 1100, 4200, 800, 3000, 1800, 2500, 2500, 1700, 3000, 2900, 2100, 2300, 2500, 1500, 3500, 2600, 1300, 2100, 3600

Table 7.3 Frequency distribution of experimental data

Cost interval	Midpoint	Frequency	Cumulative frequency
750–1250	1000	2	2
1250–1750	1500	3	5
1750–2250	2000	3	8
2250–2750	2500	5	13
2750–3250	3000	4	17
3250–3750	3500	2	19
3750–4250	4000	1	20
Total		20	

Table 7.3 shows the tabulation of the cost data as a frequency distribution. Note how the end points of the class intervals are selected in such a way that no recorded data point falls at an end point of a class. Note also that seven class intervals seem to be the most appropriate size for this particular set of observations. Each class interval has a spread of 500 which is an approximation obtained from the following expression.

$$W = \frac{X_{max} - X_{min}}{N}$$

$$= \frac{4200 - 800}{7}$$

$$= 485.71$$

$$= 500$$

Table 7.4 shows the relative frequency distribution. The relative frequency of any class is the proportion of the total observations which fall into that class. It is obtained by dividing the frequency of the class by the total number of observations. The relative frequency of all the classes should add up to 1.

From the relative frequency table, it is seen that 25% of the observed process costs fall within the range of 2250 and 2750. It is also noted that only 15% (0.10 + 0.05) of the observed process costs fall in the upper two intervals of process costs. Figure 7.7 shows the histogram of the frequency distribution for the process cost data. Figure 7.8 presents a plot of the relative frequency of the process cost data. The plot of the cumulative relative frequency is superimposed on the relative frequency plot. The relative frequency of the observations in each class represents the probability

Table 7.4 Relative frequency distribution of process cost data

Cost interval	Midpoint	Frequency	Cumulative frequency
750–1250	1000	0.10	0.10
1250–1750	1500	0.15	0.25
1750–2250	2000	0.15	0.40
2250–2750	2500	0.25	0.65
2750–3250	3000	0.20	0.85
3250–3750	3500	0.10	0.95
3750–4250	4000	0.05	1.00
Total		1.00	

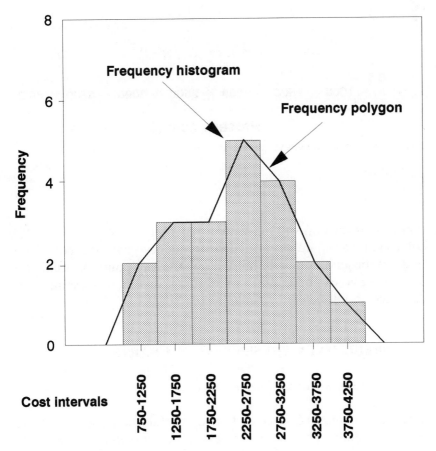

Figure 7.7 Histogram of process cost distribution data.

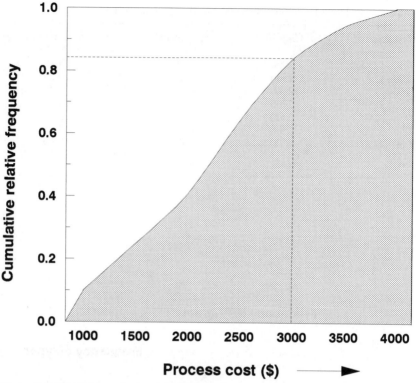

Figure 7.8 Plot of cumulative relative frequency.

that an observation will fall within that range of observations. The corresponding cumulative relative frequency gives the probability that an observation will fall below the midpoint of that class interval. For example, 85% of observations in this example are expected to fall below or equal 3000.

7.6 USEFUL DATA PRESENTATION CHARTS

A picture is worth a thousand words.

This section presents a collection of useful chart formats for representing data. The charts are easy to develop and they succinctly convey information obtained from research data analysis.

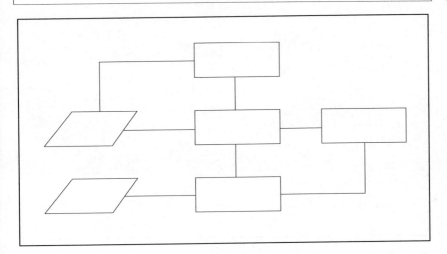

Figure 7.9 Process flowchart: shows the steps and flows of operations in process.

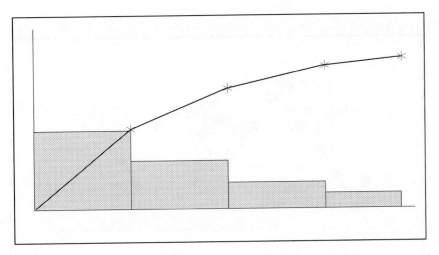

Figure 7.10 Pareto chart: identifies the most frequent causes of problems and where to focus improvement efforts. It uses ordinal and nominal data.

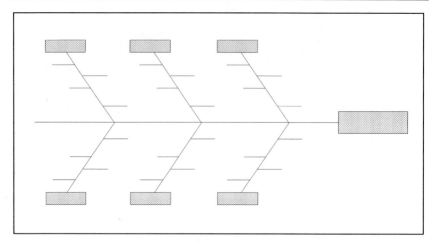

Figure 7.11 Fishbone diagram: this is a cause and effect diagram that illustrates relationships between causes and problems. It subdivides causes into their underlying components. It uses nominal data.

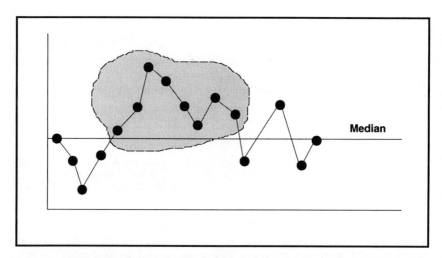

Figure 7.12 Run chart: shows changes in process over time. It indicates number of data points below and above a predetermined median level. Data include time and measure or count.

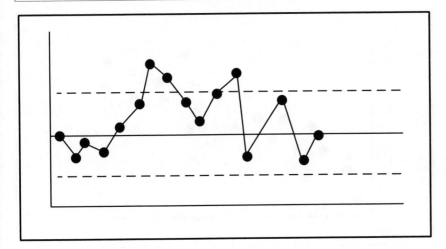

Figure 7.13 Control chart: detects change in a process and monitors the performance of the process over time. Data include time and measure or count.

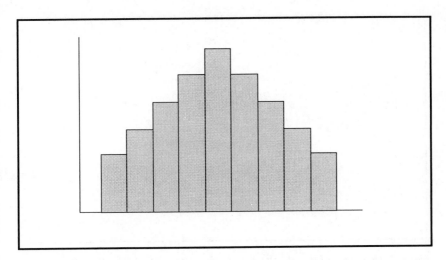

Figure 7.14 Frequency histogram: measures frequency in each data category; uses interval data.

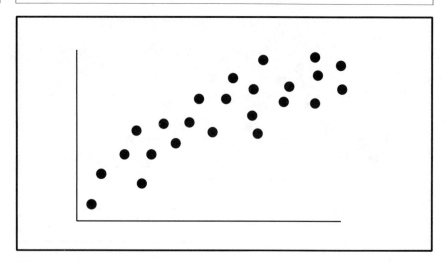

Figure 7.15 Scatter plot: illustrates the relationship between two sets of data. Uses ordinal and interval data.

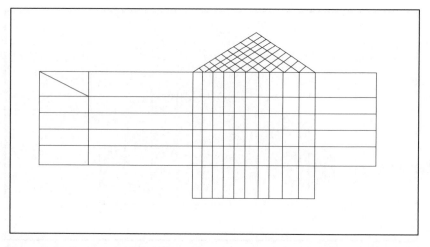

Figure 7.16 House of quality chart: illustrates the relationships between customer requirements and product design.

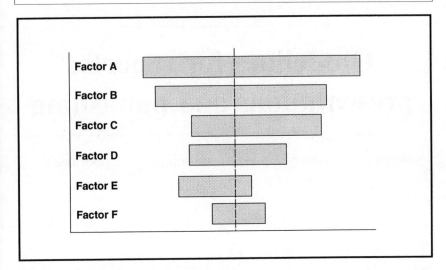

Figure 7.17 Tornado chart: expresses the effect of the range of uncertainty for each category of decision factor. Uses interval data.

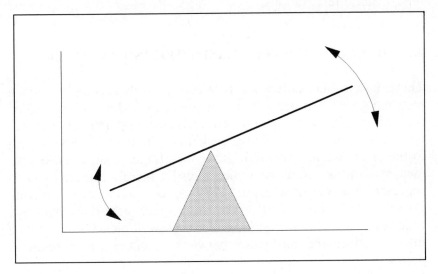

Figure 7.18 See-saw chart: illustrates relative weight or sensitivity of one factor with respect to another factor. Uses interval or ratio data.

8 | Guidelines for reports, presentations and publishing

A good idea poorly expressed will sound like a poor one.

The quality of the presentation of your research results, both in writing and orally, will determine how the audience perceives the work. This chapter presents guidelines for effective technical reports and presentations.

8.1 ORDINARY REPORT VERSUS TECHNICAL PAPER

There is a subtle difference between an ordinary report and a research paper. An ordinary report involves finding and recording facts already available somewhere. A research paper involves not only finding the facts, but also evaluating, interpreting and adding to the facts. While a research paper may require you to have your own hypothesis about the topic, a report merely requires you to represent the facts. A report may be comprehensive, covering various related ideas and topics. A research paper, by contrast, requires a specific and sufficiently narrow topic. The following titles illustrate the difference between a report and a research paper:

- Report title: 'Applications of learning curves in business'
- Research paper title: 'Extensions to the log-linear learning curve for cost control'
- Report title: 'Project management in industry'
- Research paper title: 'Resource-constrained scheduling heuristics in multi-product assembly lines'

A common mistake by students is to want to put everything imaginable into their reports. Many items contained in student reports are irrelevant 'fillers'. Part of the training of graduate students should involve the development of the maturity to be concise in technical reports. Putting too much in a write-up is often a sign of lack of confidence that an adequate job has been done. The steps to a research paper are presented below:

1. Choose the right topic.
2. Develop a succinct, representative and informative title.
3. Collect and evaluate supporting materials.
4. Organize ideas.
5. Develop outline of paper in accordance with paper specifications.
6. Write, edit, refine, proofread.

8.2 GUIDELINES FOR TECHNICAL REPORTS

In general, your report should cover the background, objective, methodology, results, and contribution. Improve both the aesthetics and contents of the report by following the guidelines below.

8.2.1 Report format

1. You should have a cover page clearly showing title, your name, date and topic.
2. Include an abstract that describes the problem, methodology, results and conclusion.
3. Follow any specific requirements for the report. For example, most reports require double spaced abstract and text, single spaced reference list with a space between entries.
4. Journal article references should contain name of author(s), year of publication, article title (in quotes), journal name (in italics, boldface, or underlined), volume number, issue number and inclusive pages. Use any of the standard reference formats available and be consistent.
5. For long reports, include a table of contents.
6. The first two sections typically should cover an introduction and a literature review. The introduction should describe the gen-

eral background for the problem. A subsection of it may present an explicit problem statement. The literature review should discuss what has been done before (as documented in the literature) and how your work differs from or adds to the existing work.

7. Always paginate your report. The preliminary pages should be numbered with lower-case roman numerals.

8.2.2 Stages of the report

1. Select appropriate topic.
2. Develop rough outline.
3. Conduct literature survey.
4. Develop rough draft.
5. Perform editing.
6. Develop final draft.

8.2.3 Progress report

If the report is a progress report, preliminary report, or an intermediate report, the contents should be organized as follows:

1. Background.
2. Objective.
3. Methodology:
   (a) updated literature review (if any);
   (b) revision in methodology;
   (c) innovative approach.
4. Results to date:
   (a) work accomplished since previous report;
   (b) relevance to previous work;
   (c) continuation from previous work.
5. Problems encountered:
   (a) problems with methodology;
   (b) problems with research tools (software, hardware, equipment);
   (c) how problems were addressed.
6. Proposed work and timeline:
   (a) new ideas;
   (b) activities and time estimates;
   (c) proposed directions.
7. Expected contribution to overall research agenda.

8.2.4 Use of figures and tables

- Use graphical and tabular presentation of information whenever possible.
- Label all figures and tables. Use concise and descriptive captions for the figures and tables.
- Each figure or table should be referenced prior to (but close to) its location in the text.
- When a figure or table is copied from another source, clearly make reference to the source.

8.2.5 Use of appendices

- Use appendices to elaborate the organization of your report.
- All incidental or lengthy materials not appropriate for the body of the text should be placed in appendices. These include mathematical derivation, lengthy proofs, computer printouts, flowcharts and experimental data. The conclusions derived from these materials should be included in the body of the text.
- Each appendix should have a title and be referenced in the body of the text. Label appendices using upper case letters, A, B, C and so on.
- Appendices should be placed at the end of the report after the list of references.

8.2.6 Use of computer materials

Most science and engineering research nowadays makes use of computer tools.

- Use computer tools effectively. Do not throw in computer analysis indiscriminately.
- Trim and burst outputs printed on computer paper before including in reports.
- Computer outputs should be reduced to the standard paper size used for the report (e.g. 8.5 by 11) but should still be readable.
- Document computer code with flowcharts.
- Use comments within computer code to document the logic used.
- Refer to and observe standard coding practices that increase readability of computer code.

8.2.7 Miscellaneous report attributes

- Pay attention to the overall quality of the report in terms of appearance and contents.
- There is no excuse for having spelling errors in your reports. Use the spell checker in your word processor.
- Proofread your report before submitting it. Allow yourself enough lead time for this 'unpleasant' part of report writing.
- Avoid abbreviations such as 'thru' for through, 'etc.' for and so on, and 'pls' for please within sentence structures. Avoid common spelling confusion such as between 'principle' and 'principal'.
- Place two spaces after each period ending a sentence, two spaces after colons, one space after commas and semi-colons.
- Avoid the temptation of using incorrect plurals such as notations, informations, equipments and softwares.
- Recognize proper singular/plural forms of commonly used words: datum/data, criterion/criteria, phenomenon/phenomena, medium/media. For example, use 'data are available' instead of 'data is available'.
- Avoid the use of first person singular 'I' and use the plural 'we' sparingly.
- Spell out numbers up to and including ten.
- Look over the report to ascertain its structure and first-look appearance.
- Avoid plagiarism. Never use materials from other sources without proper attribution. Cite your references, credit your sources, and use quotation marks as appropriate. Remember the following points:
  - (a) all summaries, quotations, and ideas borrowed from other sources must be documented;
  - (b) every source used in the paper should be included in the list of references;
  - (c) quoted materials should be placed within quotation marks or in indented block form to clearly identify them as works from other sources.

8.2.8 Writing diagnostics

When students write, they often have problems in several areas. The following are some of the most common problem areas.

- Coming up with ideas: this can be solved by doing a thorough literature search. Reflect on the ideas presented in the literature and try to come up with your own ideas to complement or refute what you read.
- Finding data to support the paper: again, a literature search will help here. Pre-writing the paper with hypothetical data will also help you identify what sort of data you need to search for.
- Coming up with a narrow and focused thesis statement: consult with the research advisor for help.
- Giving examples: develop enough relevant examples, but avoid giving too many. Recall the logic and premise of the paper as you find suitable examples.
- Organization of the materials: start with a broad outline and iteratively rearrange the outline as you expand to detailed contents.
- Proofreading: never compromise proofreading. If possible ask others to read the paper as well. Check cohesion and coherence of the sections of the paper. Check for spelling, punctuation, and grammatical errors. Do sentences always say what you mean? Is the language and format appropriate for your audience?

8.3 GUIDELINES FOR TECHNICAL PRESENTATIONS

'Tell me and I will forget; show me and I may remember; involve me and I will understand.' Chinese proverb

Research steps usually encompass the following:

- formulation of the problem
- analysis of the problem
- search for alternate solutions
- selection of best solution methodology
- presentation/defense of the solution

Presentation is the last crucial part of the research effort. Sufficient detail must be provided to fit the needs of the audience being addressed. The higher in the hierarchy the audience, the lower the level of detail should be. Since your research advisor is essentially your research collaborator, he or she must be apprised of the full details of your work. The presentation should proceed via the three steps below:

1. **Agenda**: tell the audience what you plan to tell them.
2. **Delivery**: tell the audience what you promised.
3. **Summary**: tell the audience what you presented.

8.3.1 What is research?

Research means inventing or developing a new idea and proving that the idea works. The 'proving' part of this definition of research depends on how well the idea is presented, assuming, of course, that the idea is correct to begin with.

Identify focus

- Be specific and identify the most important focus of the presentation.
- Identify important elements that support the focus.
- Describe the nature of the work in one sentence.
- Explain the methodology briefly.
- Highlight the contribution of the work.

Supporting documentation

- Prepare documentation (e.g. handouts) to back key points
- Give the audience the option of examining specific points in detail later.
- Use concise presentation of statements.
- Check spelling in the documentation.
- Documentation may use smaller typeface than presentation slides.

Identify the audience

- Who is attending (practitioners, researchers, etc.).
- Why that person is attending.
- Background of the audience.
- What the audience may already know or not know.

Audience attention

- Organize the presentation to get and keep audience attention.
- Use 'few words to many ideas' ratio approach.

- Reiterate key points.
- Elicit audience reaction or comments at key points.

Participative approach

- Use participative approach to impart the message.
- Use sight, sound, and action to make message sink in.
- Adults absorb, retain and learn according to the following progression:
  - (a) 10% of what they read;
  - (b) 20% of what they hear;
  - (c) 30% of what they read and hear;
  - (d) 50% of what they hear and see;
  - (e) 70% of what they recite to themselves;
  - (f) 90% of what they do.

Use of time

- Timing is the essence of effective presentation.
- Time is a precious commodity that must be used well.
- Use presentation time effectively.
- Be concise and thorough.
- Avoid reading from your presentation slides.
- Not too short, not too long:
  - (a) long presentations become boring and lose the audience;
  - (b) it should not take long to convince someone of a good idea;
  - (c) if it lasts more than 30 minutes, it is probably too long.

Presentation materials

- Use graphical and tabular representations whenever possible.
- Use adequately large typeface for slides.
- Do not include too much on each page (no more than ten lines).
- Keep lines short (no more than seven words); avoid full sentences.
- Use no more than five columns in a table.
- Leave adequate space in between lines.
- Use letters no less than 0.25" high.

8.4 MAJOR COMPONENTS OF TECHNICAL PRESENTATIONS

Technical presentations should progress through three stages: introduction, development and results. These three major components can be expanded to cover the elements identified below.

8.4.1 Introduction

- Should get the attention of the audience. Use a combination of the following:
  - (a) opening gag (humorous or serious);
  - (b) significant results or facts;
  - (c) conclusion that demonstrates the importance of your topic.
- Establish your credibility:
  - (a) background;
  - (b) current and/or past related job functions;
  - (c) how long the research has been going on.
- Identify focus of the presentation:
  - (a) state purpose of the presentation.
- Relate your topic to the audience:
  - (a) state why the topic is important to the field;
  - (b) state how the audience might use the results.
- State the major points (not too many) to be covered:
  - (a) what the audience can expect to learn.

8.4.2 Body of presentation

- Related works in the area:
  - (a) literature search;
  - (b) literature review;
  - (c) case study.
- Methodology:
  - (a) existing tools, models, techniques, etc.;
  - (b) what is new–background, objective, results to date, proposed work and contribution;
  - (c) variables of interest;
  - (d) theoretical, analytical, conceptual, actual, experimental, etc.
- Results.
- Validation:
  - (a) what is validated;
  - (b) basis for validation.

8.4.3 Conclusion

- Review or summarize main points of the presentation.
- Re-emphasize the results.
- Recommendations.
- Identify areas for further research.
- Refer back to introductory statements to tie everything together.

8.4.4 Recap

- Be able to say in one sentence what your research objectives are.
- Be able to state your research methodology in one sentence.
- Be able to state your contribution in one sentence.

8.4.5 Exit

- Departing gag (humorous or serious).
- Leave lasting impression on the audience.

8.4.6 Presentation style

- Practice whenever possible. This will:
    - (a) help to reduce nervous tension;
    - (b) help to maintain presentation flow;
    - (c) help identify transition points;
    - (d) help identify time requirements.
- Minimize reading from notes or note cards.
- Maintain eye contact with audience:
    - (a) shift eye contact around the room;
    - (b) don't gaze at an individual too long;
    - (c) don't leave anyone out of eye contact.
- Use gesturing (not excessively) and movements to get attention.
- Maintain appropriate and consistent vocal volume.

8.5 PRESENTATION EVALUATION FORMS

The format presented below is useful for evaluating the performance of a student at a technical presentation or defense of thesis research. The layout can be modified or altered to fit specific evaluation needs.

Candidate's Name:

Candidate for: MS   PhD

Part I: Evaluation of comprehensive oral examination

A: Rating of student's substantive knowledge

Examiner 1: . . . .

Questions	Excellent	Very good	Average	Fair	Poor
1.					
2.					
3.					
4.					

Examiner 2: . . . .

Questions	Excellent	Very good	Average	Fair	Poor
1.					
2.					
3.					
4.					

Examiner 3: . . . .

Questions	Excellent	Very good	Average	Fair	Poor
1.					
2.					
3.					
4.					

Excellent: wide and deep knowledge of subject matter and of its role in relation to his/her special area as well as to general area of science and engineering. Good knowledge of scientific principles.

Very good: detailed knowledge of subject as well as grasp of broad area. Strong awareness of relevance of facts to each other and of underlying principles and implications of results.

Average: generally accurate information and knowledge of basic facts of topic. Some appreciation of the organization of these facts in his/her own area and relevance to other areas.

Fair: minimal accurate information. Supporting evidence or data are weak. Only occasional evidence of awareness of relevance of knowledge and information to developing plans and utilization.

Poor: inaccurate and/or irrelevant statements. Misinformation. Weak grasp of principles and interrelationships between facts and research.

B: Student's Presentation Skills

Grading of oral presentation can be based on the following:

(a) Mechanics of presentation (organization, use of visual aids, effective use of time, etc.).
(b) Quality of presentation (clarity, handling questions, knowledge of the problem and solution methods, etc.).

Item (b) should carry more weight (e.g. weighted twice as much) than (a). It can be evaluated based on the ordinal scale below:

Excellent: quickly grasps and develops the point of a question. Presents logical, well-organized response. Is clear, confident and poised. Uses nice visual aids.

Very good: understands questions and gives relevant well-organized response. Appears to be in good command of him/herself and the situation.

Average: for the most part, presents replies in an organized and suitable way. Needs occasional probing to draw him/her out. Is somewhat self-confident.

Fair: often responds slowly. Asks interviewer to repeat and explain questions too often. Appears uncertain of him/herself, either by being diffident or by obviously bluffing.

Poor: has difficulty in replying to questions. Misses the point, but does not know it. Needs many probing questions to produce an answer at all. Illogical statements. Gives quick, evasive and diffuse responses.

Part II: Overall evaluation of student's performance on oral exam:

high pass
pass
minimal pass
fail

Comments and notes on recommended remedial work:
Part III: Summary of student's written paper or thesis

Title of paper:
Date submitted:
Date signed off:

Performance metric	Excellent	Very good	Average	Fair	Poor
1. Difficulty of topic					
2. Creativity in dealing with topic					
3. Demonstrated mastery in use of scientific procedures					
4. Ability to write clearly and concisely					
5. Ability to follow accepted format					
6. Absence of obvious errors, typos and grammatical mistakes					

Part IV: Final evaluation

Satisfactory
Unsatisfactory
Decision deferred

8.6 GROUP AND CLASSROOM PRESENTATION EVALUATION

In cases where technical presentations are to be made by student groups in a classroom setting the performance evaluation can be based on a peer rating approach. Each student in the class (the audience) will have an opportunity to evaluate the presentation of every other student in a round-robin presentation format. Every member of each group is expected to contribute effectively to the group project.

Each student will submit a confidential evaluation of each member in his or her group at the end of the semester. The weighted evaluation will then be used in distributing the final group grade to the members of the group. The format presented below is suggested:

Total points available to a group: $100n$ (n = number of students in the group)

Rate the contribution of each group member on a scale of 0 to 100.
Rating for student 1: x_{1j} (where j is the student doing the evaluation)
Rating for student 2: x_{2j}

Rating for student n: x_{nj}
Total rating points 100n

Weighted rating for student i is then calculated as:

$$w_i = \frac{1}{n} \sum_{j=1}^{n} x_{ij}$$

	Rating by student 1 ($j = 1$)	Rating by student 2 ($j = 2$)			Rating by Student n ($j = n$)	w_i
Rating for student i = 1						
Rating for student i = 2						
. . .						
. . .						
Rating for student n						

The overall project score for student i is calculated as:

$S_i = (w_i)(\text{grade assigned to project})$

8.7 GUIDELINES FOR PUBLISHING

Publication is the ultimate avenue for disseminating your research. A science or engineering technical paper can be one of the following three forms:

- survey paper
- experimental paper
- theoretical paper

8.8 DETERMINING APPROPRIATE MEDIA

For your publication to be effective, it must appear in an appropriate medium. There are trade magazines and there are archival journals. Most engineering and science research will be targeted for archival journal publication. Key factors to consider in determining where to publish are:

- focus and scope of the journal;
- the circulation volume of the journal;
- relevance of the journal's focus to your area of specialization;
- the nature and length of the review process;
- the lead time to publishing papers accepted for publication.

Be sure to get a copy of the author's guide for the journal where you expect to publish. Follow the guidelines as much as possible. For cases where a call for papers is issued by the journal, try and operate within the specified guidelines and deadlines. A later chapter presents a sample of a published technical paper.

8.9 USE OF APPLICATIONS SOFTWARE

Application software packages can add flexibility and functionality to your document processing needs. A complete research facility should have the different types of applications software discussed above. You should use them in such a way that they complement each other. You can use Lotus Freelance for drawing charts and figures for research presentations. The graphics files can be merged with word processing documents to produce complete publishable research reports. Spreadsheets can be used to format numeric tables of experimental data. These, like graphics, can be merged with reports. Sometimes, researchers have database files that they would like to include in their word processing documents to create integrated reports. Such database files can be retrieved with a database manager program, organized into an appropriate format, and then exported to the word processing document.

If you would like to have all your application programs right at your finger tips, so to speak, you may want to procure an integrated application software package. Such a single program may include modules for several applications programs like word processing, communication, database management, spreadsheets and graphics. An integrated package permits easy manipulation and interchange of data in several different applications. Moreover, it allows the user to learn one consistent set of commands for accomplishing a variety of tasks instead of learning the individual sets of commands for independent packages. But integrated programs can be cumbersome to install and use, particularly if you need only one of the several functions of the package. For this reason, developers are moving away from integrated packages in favor of individual but compatible packages.

Programs written by users are referred to as user-written programs. Those written by professional programmers and sold to users are referred to as pre-written application packages, canned programs or commercial packages. Shareware programs are those written by an independent source and shared with the interested user community free of charge. Sometimes voluntary donations are solicited for maintaining shareware programs. Commercial packages can be purchased from a vendor, distributor or directly from the developer. The advantages and disadvantages of commercial application packages are listed below.

8.9.1 Advantages

They:

- can be purchased for immediate use;
- can be inexpensive;
- are available for many applications;
- have already been tested;
- usually offer technical support from vendor.

8.9.2 Disadvantages

They:

- may be too general to fit specific user's needs;
- may have been developed for a computer not compatible;
- frequent updates can disrupt learning and ease of use;
- may require operating environment that is not available to user;
- can be expensive.

The advantages and disadvantages of user-written application packages are presented below.

8.9.3 Positive aspects

They:

- can be written to fit the exact needs of a user;
- can be sold to others if successful;
- give the pride of generating the computer code.

8.9.4 Negative aspects

- A high level of programming expertise is required.
- They have a long development time.
- The development process can be frustrating to the user
- There is a lack of formal testing.

The most frequent application software packages fall into five categories:

- word processors
- data managers
- electronic spreadsheets
- graphics
- communications

8.9.5 Word processor

Typewriters are quickly becoming obsolete. The proliferation of computer tools is making manual document processing a thing of the past. Even the filling out of forms is becoming computerized. A word processor is a program for manipulating text. It allows you to type a document into the microcomputer's memory. You may view the document on the screen as you type. You can move directly to any point within the document to add, delete, copy or move sections of text. You can save the document for later retrieval or you can print it. WordPerfect, Microsoft Word and Lotus Ami Pro are three of the most popular word processing packages.

8.9.6 Data manager

In many graduate research situations, you will need to deal with sets of data. Such a collection of data can be analyzed or updated by using a database management program. A database program allows you to add to the data or retrieve data using some specific criteria. Specified data characteristics may be used to determine how data elements can be organized, manipulated, processed, and presented. DBASE, RBASE 5000, ACCESS, and FoxPro are some of the most popular database programs.

8.9.7 Spreadsheet

A spreadsheet program allows you to define relationships between numbers. For example, you can define one entry as the sum of a column of numbers. As numbers in the column change, the sum is automatically updated. The program virtually turns the computer screen into a numeric work pad. Lotus 1–2–3, Excel and Quattro Pro are three popular spreadsheet programs.

8.9.8 Graphics

Graphics packages display data visually in the form of graphic images. For example, after using a spreadsheet or data manager to manipulate and organize data, it can sometimes be difficult to see the data relationships or interpret the information that is generated. Presenting the information visually (graphically in the form of pie charts, histograms, etc.) is one way to make the information clear. Researchers and analysts can use graphics packages to present statistics and other data and their relationships. Lotus Freelance, Powerpoint, and Harvard Graphics are three of the most frequently used graphics packages. Flowchart software such as ABC Flowchart and Vision are specifically designed for flowcharting purposes.

8.10 GUIDELINES FOR POSTER PRESENTATIONS

Poster presentations are often used as an avenue to disseminate research work at technical conferences or professional fairs. A poster should concisely convey the objectives, methodology, data, results and conclusions of a research effort.

8.10.1 What is a poster presentation?

A poster presentation is a technical paper presented through a variety of graphic wall or table displays. These may include maps, charts, photographs, computer outputs, scale models, and samples. The author will be in attendance during the display period. The displays make the presentation of certain types of information more effective, such as mathematical representations, graphical models, large data sets or lengthy materials. This allows the audience to review the poster displays at their own pace and to discuss the materials with the authors in detail on a one-to-one basis.

A poster session provides a unique opportunity for authors to present a technical paper, while affording the attenders an easier format for asking questions and receiving in-depth responses. Many presenters and attenders feel more comfortable in a self-paced discussion than a condensed formal presentation.

8.10.2 Materials needed

- poster board (about 4 ft by 8 ft), usually provided at the conference site;
- supplies; tacks, tape, scissors, glue, velcro materials, etc.

8.10.3 What to include

Poster title, diagrams, charts, figures, etc. These should be sized and lettered so they are legible and readable by a group of attendees at a distance of about five feet. They should also be simple, colorful (if possible), well labeled, neat and well organized. Remember, the poster should succinctly highlight:

- problem statement
- objective
- methodology
- data analysis
- results
- conclusions

8.10.4 General guidelines

- Do not merely display data. It is important to show the implications and significance of the work.
- All display materials should be easily readable at a distance of about five feet. If this appears impossible for a specific item, then handouts should be used to convey that particular information.

- Determine in advance a logical and attractive layout for your poster. If the poster is composed in sections, number the sections so that they can be easily re-mounted at the conference site.
- Find out in advance your poster session number and booth location and verify before mounting your poster at the site.
- Display the title, author and affiliation in large enough letters at the top of the poster board.
- Take with you useful items such as sketch pads and marking pens.
- Make sure your materials are sufficiently lightweight and thin to be mounted on a vertical board. For heavy and thick materials, consider using velcro material (available in fabric stores) for mounting the items.
- Do not write on display boards.
- Attempt to have your poster completely set up at least 15 minutes before your presentation. Take your time to remove all your materials from the poster board at the conclusion of the display.
- During the presentation, discuss your topic conversationally rather than lecturing or simply reading a summary. The discussion may begin with a question from an interested attendee. You may initiate a discussion by pointing out the figure that depicts the conclusions of your research. Questions and explanations can then follow from that point.

8.10.5 Handouts

It is a good idea to have handouts to supplement the poster displays. Posting of the entire paper on the poster board is not recommended. Members of the audience always appreciate receiving copies of handouts. A handout gives an attendee something to look forward to (in terms of reading) about your research. The impression of your research will probably last longer with someone who has received a copy of your handout than someone who has not. Remember, a combination of auditory and visual aids is the best way to convey your ideas.

8.10.6 Poster evaluation criteria

A poster session can be rated on an ordinal scale from excellent to inadequate (excellent, good, fair, poor, inadequate). Some of the specific criteria are:

- Poster is well designed, organized and attractive.
- Major points of the work are presented.
- Diagrams present enough materials to clearly identify:
  - (a) problem
  - (b) objective
  - (c) methodology
  - (d) data analysis
  - (e) results
  - (f) conclusions
- Oral presentation/explanations are clear and understandable.
- Handout materials are available and useful.

Electronic communication | 9

Not all that is fit to print is fit to read.

9.1 EMERGENCE OF ELECTRONIC COMMUNICATION

You can use communications programs to access information in distant computers. This is sometimes referred to as remote access. As more individuals and organizations use computers, the need to transfer data from one computer to another has increased. For example, law enforcement officials exchange information on criminals, home users can access stock market information, or businesses can exchange sales information via a remote computer linkage. Electronic mail (e-mail) is another useful application of computer communications. Electronic mail allows multiple access communication that is carried out exclusively on a computer network.

Commercial services such as MCI Mail, CompuServe and Western Union's Easylink provide electronic 'mailboxes' into which people can deposit mail. Using electronic mail, you can send or receive memos, reports, letters and other documents to and from anywhere in the world. Word processing service centers can, for example, deliver their clients' completed manuscripts via electronic mail. This, of course, assumes that the client has compatible hardware and software to generate a hardcopy print out of the manuscript. Some of the popular communications packages are Cross-Talk, PC-Talk and ProComm.

9.2 THE INTERNET

Internet may be defined as a network of networks. It is a worldwide network that connects computers. It is the world's largest computer network. The proliferation of desktop computers and the communication revolution have led to the emergence of the information superhighway. This offers new opportunities to communicate and facilitate information exchange. The introduction of electronic journals to disseminate scholarly information is now a major effort being pursued by many professional organizations. An electronic journal will serve as an interactive medium to deliver timely and updated information to researchers. The use of Internet will certainly continue to grow exponentially as new tools and techniques are developed to facilitate online communication.

Internet initially served the academic community. It now serves the business world as well. It provides low-cost global communication. Several hundred specialist databases can be accessed through Internet. Internet electronic mail (e-mail) is now the most efficient and speedy way to communicate. Users can send messages to groups with a single command. Companies can receive information from experts in especially narrow areas of research. With Internet, researchers can conduct collaborative research with very little time lag between communication. One very useful aspect of Internet is that a variety of applications software packages are available free and can be downloaded.

Most of the local area networks (LAN) available at academic institutions are connected to Internet. If you are connected to Bitnet or Telnet, then you are on an Internet node. Some of the advantages of Internet and the associated e-mail facilities are:

- ability to send and receive messages quickly;
- ability to send the same message to a large group of people with a single command;
- ability to provide paperless communication (this is good news to save-the-tree groups around the world);
- ability to link to remote locations through wireless technologies;
- ability to have a history of communication actions;
- ability to access important databases that cannot be physically accessed;
- ability to always be in touch (no more excuses about missed phone calls as long as access to a computer network is available).

9.2.1 The Worldwide Web

Worldwide Web refers to the web-like connection of computer networks. It is a distributed hypermedia environment involving a large, informal and international design and development team. It is a universe of information woven together from several related networks. The elements of the web include text and multimedia documents, the rules and tools that link the documents, and the addresses (pathways) that link the documents. The addresses allow a user to browse documents available in the Worldwide Web. The searching and browsing of Web documents is often called surfing.

The Worldwide Web originated from the European Particle Physics Laboratory in 1989. The laboratory was formed by a collection of European high-energy physics researchers. The original idea (proposal) involved the development of a means for disseminating research and ideas effectively throughout the organization. The first piece of Web software was introduced in 1990. It had capabilities for viewing hypertext documents and transmitting them to other people on the Internet.

With the passage of time, developers added to the capabilities of the Worldwide Web. The developers created the protocol, which is the set of rules and specifications used by different computer networks to communicate with one another. The protocol makes it possible for tools such as Mosaic to communicate with computers that store Web documents. In 1993 the popularity of Web exploded across the Internet. The introduction of Mosaic was credited with the exponential growth.

9.3 USE OF GOPHER

Gopher is an information service that uses Internet to connect systems and the information service they provide all over the world. Gopher is becoming increasingly popular around the world. Gopher knows the connection protocols and connects you automatically to the menu item you select. Each Gopher server puts information and/or services on a public access menu that any other Gopher server can access. Menus are interconnected in a web-like structure. There may be several different ways to reach any given destination.

Information available on Gopher may include telephone directories, job listings, on-line journals, reference services, catalogs and special interest professional networks. Any information available to a group on any Gopher server is also available to the whole world. There are presently active Gopher servers on six continents and in several nations. Gopher was developed at the University of Minnesota. It is said that it was so named because it is supposed to 'go for' information but others claim that it was named after the Minnesota team, the Golden Gophers.

Gopher is organized in menu trees. Your first view of Gopher will be the local menu, from which you can select an item. Gopher gets the next level of information, which may be another menu or may be text or other information. The data you want to look at may be several menus deep and may actually be located in a distant (farther than remote) location. The menus point you via Internet to the information. You may never know exactly which computer at what location or in which country is providing the information or the connections for your use.

Gopher is a global network of interconnected communication networks. Links exist between gophers. Such links may connect you to a gopher in Oklahoma, New York, Nigeria, London, Taiwan, Moscow, Mexico or Finland. Not all the systems and links are in operation at the same time all over the world. Consequently, there may be times when you may not be able to have desired access. If you experience access problems at any particular time, simply try at a later time. The Gopher service may be effectively used to secure information relevant to your research work.

There are many ways to get files and receive mail online. The easiest way is to get an account with an online information service such as CompuServe or America Online. A direct Internet connection gives you access to everything available on the network. Commercial online systems own the network and the information available on it and they may charge access or connection fee for the information.

9.4 USE OF MOSAIC

Mosaic is an Internet-based global hypermedia browser that allows you to discover, retrieve and display documents and data from all over the Internet and around the world. It is a graphical Worldwide Web user client application program. A browser is a utility tool that displays hypertext files. Mosaic was developed at the National Center for Supercomputing Applications (NCSA) at the University of Illinois at Urbana-Champaign. The Mosaic project was funded by the US National Science Foundation, a government agency.

Because of Mosaic's public funding, it is available free of charge to any interested person. It comes in three options: Microsoft Windows, X Windows (UNIX) and Macintosh.

Mosaic is a 'point-and-click' windows-type graphical interface to all of the resources of the Internet. Mosaic provides Internet access without direct communications and connection complexity. Commercial, government and educational 'Home Pages' offer instant access to the world's information superhighway using hypertext and hypermedia. New Mosaic sites are being established as individuals, businesses, government agencies, universities and many other types of organization join the cyberspace. The newer version of Mosaic is called NetScape.

The Web is a client–server networking system, just like the Net in Internet. When you use Mosaic to request information from another computer on the Net, your computer and Mosaic act as the client. The computer that responds to your request is a server. A server can be accessed by a variety of users who have client programs that are supported by the server.

Conventional Internet is a text-only tool in a UNIX environment. Mosaic enhances this environment to a communications rich multimedia environment of hypertext links, graphics, sounds and videos. Mosaic lets you achieve the following:

- Work on the Internet using common Windows menu commands, buttons, and dialog boxes as well as your computer's mouse.
- Display electronic documents with text in a variety of formats, including numbered and bulleted lists, paragraphs, and numerous fonts in normal, bold and italic styles.
- Connect to programs that let you view pictures, play audio sounds, and watch videos in a wide variety of file formats.
- Use interactive, electronic forms that include such easy-to-use elements as fields, check boxes and radio buttons. For example, you might fill out a generic application form for admission into a graduate program and send it to multiple schools all at once.
- Use as many as 256 colors for formatting documents and displaying pictures and graphs.
- Create customized menus that let you automatically connect to your favorite destinations.
- Download copies of Web documents as files to your computer, where you can use Mosaic to work with them locally without a live connection to the Internet.

To take advantage of the Mosaic developments for graduate research purposes, these are some of the things you need to do:

- Understand the jargon of the Worldwide Web, Mosaic and the Internet.
- Obtain a copy of Mosaic for Windows, which can be downloaded free to your computer.
- Get the right Internet connection to run Mosaic.
- Use Mosaic to work with Gopher, FTP and other Internet tools.
- Find hundreds of Worldwide Web sites for information related to your research.
- Create your own Web documents and make them available to millions of Internet users.

As a part of the supporting skills for your graduate research, you should become familiar with the e-mail and other communication network facilities available to you. These can be effectively used in communicating with other researchers to obtain relevant information for your research. Summary guidelines for e-mail are presented below.

9.5 GUIDELINES FOR E-MAIL

Many in-house computer systems provide an electronic mail (e-mail) service for communicating with other computer users. From time to time when you log in, you will see the message You have mail before you get the systems prompt. This means that you have mail waiting for you. You can then access the mail by going through the prescribed steps in accordance with the specific setup of your computer system. Selected e-mail commands and their functions are described below:

- **Command name**: mail
- **Syntax**:

  mail [-v] [-i] [-n] [-e] [-s subject] [user ...]
  mail [-v] [-i] [-n] -f [name]
  mail [-v] [-i] [-n] -u user
  mail nodename::username (if node is installed)

- **Description**: the mail utility is an intelligent mail processing system which has a command syntax similar to an editor. However, in mail lines are replaced by messages.

- **Sending mail**: to send a message to one or more persons, type mail and the names of the people receiving your mail. Press the RETURN key. Note that if you use other arguments, the names of the recipients should always be the last element on the command line. For example:

  mail -v -s 'mail message' users

If you do not specify a subject on the command line, you are prompted for a subject. After entering a subject, and pressing the RETURN key, type your message. To send the message, type a period (.) or CTRL D at the beginning of a new line. You can use tilde (~) escape sequences to perform special functions when composing mail messages. See the list of options for more on tilde escape sequences.

- **Reading mail**: in normal usage mail is given no arguments and checks your mail out of the mail directory. Then it prints out a one line header of each message there. The current message is initially the first message and is numbered 1. It can be displayed using the print command. The -e option causes mail not to be printed. Instead, an exit value is returned. You can move among the messages by typing a plus sign (+) followed by a number to move forward that many messages, or a minus sign (−) followed by a number to move backward that many messages.

- **Disposing of mail**: after reading a message you can delete (d) it or reply (r) to it. Deleted messages can be undeleted, however, in one of two ways: you can use the undelete (u) command and the number of the message, or you can end the mail session with the exit (x) command. Note that if you end a session with the quit (q) command, you cannot retrieve deleted messages.

- **Specifying messages**: commands such as print and delete can be given a list of message numbers as arguments. Thus, the command delete 1 2 deletes messages 1 and 2, while the command delete 1–5 deletes messages 1 to 5. The asterisk (*) addresses all messages, and the dollar sign ($) addresses the last message. For example, the top command, which prints the first few lines of a message, can be used in the following manner to print the first few lines of all messages.

- **Replying to or originating mail**: use the reply command to respond to a message.

- **Ending a mail processing session**: end a mail session with the quit (q) command. Unless they were deleted, messages that you have read go to your mbox file. Unread messages go back to the mail directory. The -f option causes mail to read in the contents of your mbox (or the specified file) for processing. When you quit, the mail utility writes undeleted messages back to this file. The -u flag is a short way of specifying:

        mail -f /usr/spool/mail/user

- **Personal and system wide distribution lists**: you can create a personal distribution list that directs mail to a group of people. Such lists can be defined by placing a line similar to the following in the .mailrc file in your home directory:

        alias nsfgroup sieger badiru milan marcus abidemi@mailhost

Nsfgroup is the name of the distribution list that consists of the following users: sieger, badiru, milan, marcus, and abidemi@mailhost. A list of current aliases can be displayed with the alias (a) command in mail.

- **System-wide distribution lists** can be created by editing /usr/lib/ aliases. The syntax of system-wide lists differs from that of personally defined aliases. Personal aliases are expanded in mail you send. When a recipient on a personally defined mailing list uses the reply (r) option, the entire mailing list receives the response automatically. System-wide aliases are not expanded when the mail is sent, but any reply returned to the machine will have the system-wide alias expanded as all mail goes through sendmail. Forwarding is also a form of aliasing. A .forward file can be set up in a user's home directory. Mail for that user is then redirected to the list of addresses in the .forward file.
- **Options**:
    - -e Causes mail not to be printed. Instead, an exit value is returned.
    - -f Causes mail to read in the contents of your mbox file (or another file you specify) for processing.
    - -i Causes tty interrupt signals to be ignored. This is useful when using mail on noisy phone lines.
    - -n Inhibits the reading of /usr/lib/Mail.rc.
    - -s Specifies a subject on the command line. Note that only the first argument after the -s flag is used as a subject and that you must enclose subjects containing spaces in quotes.

-u Specifies a shorthand for expressing the following:

mail -f /usr/spool/mail/user

-v Prints the mail message. The details of delivery are displayed on the user's terminal.

The following options can be set in the .mailrc file to alter the operation of the mail command. Each command is typed on a line by itself and may take arguments following the command word and the command abbreviation. For commands that take message lists as arguments, if no message list is given, then the next message forward which satisfies the command's requirements is used. If there are no messages forward of the current message, the search proceeds backwards. If there are no good messages at all, mail cancels the command, displaying the message: No applicable messages.

- Prints out the previous message. If given a numeric argument n, prints nth previous message.

? Prints a brief summary of commands.

! Executes the ULTRIX shell command which follows.

alias (a) Prints out all currently defined aliases, if given without arguments. With one argument, prints out that alias. With more than one argument, creates a new or changes an old alias. These aliases are in effect for the current mail session only.

alternates (alt) Informs mail that you have several valid addresses. The alternates command is useful if you have accounts on more than one machine. When you reply to messages, mail does not send a copy of the message to any of the addresses listed on the alternates list. If the alternates command is given with no argument, the current set of alternative names is displayed.

chdir (ch) Changes the user's working directory to that specified. If no directory is given, the chdir command changes to the user's login directory.

copy (co) Takes a message list and file name and appends each message to the end of the file. The copy command functions in the same way as the save command, except that it does not mark the messages that you copy for deletion when you quit.

delete (d) Takes a list of messages as argument and marks them all as deleted. Deleted messages are not saved in mbox, nor are they available for most other commands.

dp (or dt) Deletes the current message and prints the next message. If there is no next message, mail returns a message: at EOF edit (e) Takes a list of messages and points the text editor at each one in turn. On return from the editor, the message is read back in.

exit (ex or x) Returns to the shell without modifying the user's system mailbox, mbox file, or edit file in -f.

file (fi) Switches to a new mail file or folder. If no arguments are given, it tells you which file you are currently reading. If you give it an argument, it writes out changes (such as deletions) you have made in the current file and reads in the new file. Some special conventions are recognized for the name. A pound sign # indicates the previous file, a percent sign % indicates your system mailbox, %user indicates the user's system mailbox, an ampersand & indicates your ~/mbox file, and +folder indicates a file in your folder directory.

folders List the names of the folders in your folder directory.

folder (fo) Switches to a new mail file or folder. The folder command functions in the same way as the file command.

from (f) Takes a list of messages and prints their message headers in the order that they appear in the mail directory, not in the order given in the list.

headers (h) Lists the current range of headers, which is a 20 message group. If a plus sign (+) is given as an argument, then the next message group is printed. If a minus sign (−) is given as an argument, the previous message group is printed.

help Prints a brief summary of commands. Synonymous with ?.

hold (ho), also preserve Takes a message list and marks each message in it to be saved in the user's system mailbox instead of in mbox. The hold command does not override the delete command.

ignore Adds the list of header fields named to the ignored list. Header fields in the ignore list are not printed on your terminal when you print a message. This command is frequently used to suppress certain machine-generated header fields. The type and print commands are used to print a message in its entirety, including ignored fields. If ignore is executed with no arguments, it lists the current set of ignored fields.

mail (m) Takes login names and distribution group names as arguments and sends mail to those people.

mbox Indicates that a list of messages should be sent to mbox in your home directory when you quit. This is the default action for messages if you did not set the hold option.

next (n, + or CR) Goes to the next message in sequence and types it. With an argument list, it types the next matching message.

preserve (pre) Takes a message list and marks each message in it to be saved in the user's system mailbox instead of in mbox. Synonymous with the hold command.

print (p) Takes a message list and types out each message on the user's terminal, without printing any specified ignored fields.

Print (P) Prints a message in its entirety, including specified ignored fields.

quit (q) Terminates the session. All undeleted, unsaved messages are saved in the user's mbox file in his or her login directory; all messages marked with hold or preserve or that were never referenced are saved in the system mailbox; and all other messages are removed from the system mailbox. If new mail arrives during the session, the user receives the message: You have new mail. If given while editing a mailbox file with the -f flag, then the edit file is rewritten. A return to the Shell is effected, unless the rewrite of the edit file fails, in which case the user can escape with the exit command.

reply (r) Takes a message list and sends mail to the sender and all recipients of the specified message. The default message must not be deleted.

Reply (R) Replies to originator of the message. Does not reply to other recipients of the original message.

respond save (s) Takes a message list and sends mail to the sender and all recipients of the specified message. Synonymous with reply.

save (s) Takes a message list and a file name and appends each message to the end of the file. The messages are saved in the order in which they appear in the mail directory, not in the order given in the message list. The filename, which is enclosed in quotes, followed by the line count and character count, is displayed on the user's terminal.

set (se) Prints all variable values when no arguments are given. Otherwise, the set command sets the specified option. Arguments take the form:

    option = value or option

shell (sh) Invokes an interactive version of the shell.

size Takes a message list and prints out the size (in characters) of each message. The size of the messages are printed in the order that they appear in the mail directory, not in the order given in the list.

source (so) Reads mail commands from a file.

top Takes a message list and prints the top few lines of each. The number of lines printed is controlled by the variable toplines and defaults to five.

type (t) Takes a message list and types out each message on the user's terminal, without printing any specified ignored fields. Synonymous with print.

Type (T) Prints a message in its entirety, including specified ignored fields. Synonymous with Print.

unalias Takes a list of names defined by alias commands and cancels the list of users. The group names no longer have any significance.

undelete (u) Takes a message list and marks each one as not being deleted.

unset Takes a list of option names and discards their remembered values; the inverse of set.

visual (v) Takes a message list and invokes the display editor on each message.

write (w) Takes a message list and a file name and appends each message to the end of the file. Synonymous with save.

xit (x) Returns to the shell without modifying the user's system mbox , or edit file in -f. Synonymous with exit.

z Presents message headers in windowfuls as described under the headers command. You can move forward to the next window with the z command. You can also move to the previous window by using z-.

The following is a summary of the tilde escape functions that you can use when composing mail messages. Note that you can only invoke these functions from within the body of a mail message and that the sequences are only executed if they are placed at the beginning of lines.

~!command Executes the indicated shell command, then returns to the message.

~? Prints a brief summary of tilde commands.

~: Executes the mail commands. (For example, the command ~:10 prints out message number 10 while ~:- prints out the previous message.)

~c name... Adds the given names to the list of carbon copy recipients.

~d Reads the file named dead.letter from your home directory into the message.

~e Invokes the text editor on the message you are typing. After the editing session is finished, you may continue appending text to the message.

~f messages Reads the named messages into the message being sent. If no messages are specified, reads in the current message.

~h Edits the message header fields by typing each one in turn and allowing the user to append text to the end or to modify the field by using the current terminal erase and kill characters.

~m messages Reads the named messages into the message being sent, shifted one tab space to the right. If no messages are specified, reads the current message.

~p Prints the message on your terminal, prefaced by the message header fields.

~q Aborts the message being sent, copying the message to dead.letter in your home directory if the save option is set.

~r filename Reads the named file into the message.

~s string Causes the named string to become the current subject field.

~t name... Adds the given names to the direct recipient list.

~v Invokes an alternate editor (defined by the VISUAL option) on the message. Usually, the alternate editor is a screen editor. After you quit the editor, you can resume appending text to the end of your message.

~w filename Writes the message onto the named file.

-|command Pipes the message through the command as a filter. If the command gives no output or terminates abnormally, retains the original text of the message. The command fmt(1) is often used as command to rejustify the message.

~~ Inserts the string of text in the message prefaced by a single tilde (~). If you have changed the escape character, then you should double that character in order to send it.

Options are controlled via the set and unset commands. Options may be either binary or string. If they are binary you should see whether or not they are set; if they are string it is the actual value that is of interest.

The binary options include the following:

append Causes messages saved in mbox to be appended rather than prepended.

ask Causes mail to prompt you for the subject of each message you send. If you simply respond with a new line, no subject field is sent.

askcc Asks you at the end of each message whether you want to send a carbon copy of the message to additional recipients. Responding with a new line indicates your satisfaction with the current list.

autoprint Causes the delete command to behave like dp; thus, after deleting a message, the next one is typed automatically.

debug Causes mail to output information useful for debugging mail. Setting the binary option debug is the same as specifying -d on the command line.

dot Causes mail to interpret a period alone on a line as the terminator of a message you are sending.

ignore Causes interrupt signals from your terminal to be ignored and echoed as at signs @.

ignoreeof Causes mail to refuse to accept a control -d as the end of a message.

msgprompt Prompts you for the message text and indicates how to terminate the message.

metoo Includes the sender in the distribution group receiving a mail message.

nosave Prevents mail from copying aborted messages into the dead.letter file in your home directory.

quite Suppresses the printing of the version when first invoked.

verbose Displays the details of each message's delivery on the user's terminal. Setting the verbose option is the same as typing -v on the command line.

The string options include the following:

EDITOR Pathname of the text editor to use in the edit command and ~e escape. If not defined, then a default editor is used.

SHELL Pathname of the shell to use in the ! command and the ~! escape. A default shell is used if this option is not defined.

VISUAL Pathname of the text editor to use in the visual command and ~v escape.

crt Threshold to determine how long a message must be before the command more is used to read it.

escape The first character of this option gives the character to use in the place of tilde (~) to denote escapes, if defined.

folder Directory name to use for storing folders of messages. If this name begins with a backslash (/) mail considers it an absolute pathname; otherwise, the folder directory is found relative to your home directory.

record Pathname of the file used to record all outgoing mail. If it is not defined, then outgoing mail is not saved.

toplines The number of lines of a message that is printed out with the top command; normally, the first five lines are printed.

Return values If mail is invoked with the -e option, the following exit values are returned: 0 the user has mail, 1 the user has no mail

10 | Completion, termination and transition

Research is not work until it works.

10.1 TRANSITION TO THE REAL WORLD

What is your life after school going to look like? Completing your science or engineering graduate studies and moving on to a career is one of the many processes you will go through. Be positive and enthusiastic about the future. If you are prepared and take advantage of available opportunities, you should have a rosy professional career.

10.2 PLANNING FOR COMPLETION

You should be prepared for the completion of your graduate education just as much as you were at the beginning of the program. Good preparation will prevent unpleasant surprises. Even though you have completed the formal component of your graduate studies, you should not stop learning. An elderly university graduate once remarked that 'The world can't get stale if you continue to learn.' You will need to learn new things to support your professional goals and retain your job status.

10.3 TRANSITION FROM STUDENT TO PROFESSIONAL

Making the transition from student to professional can be stressful. Good preparation must be made. The first step of the transition is to find suitable employment. Presented below are general guidelines for your job search. The guidelines are applicable to academic and non-academic types of employment.

- Start early on your job search process.
- Prepare a comprehensive application packet that can be easily modified or adapted to the requirements of different employers. The purpose of the packet is to tell the prospective employer everything of interest about you (education, experience, social interactions, service, etc.). It must be appealing enough to motivate employers to want to interview you.
- Make your curriculum vitae (resumé) the lead element of the packet. This should not appear like a generic 'sendto-all' document. Although there are many standardized resumé formats available, you should not be tempted to use them as they are. Try to customize a selected format to yourself with your own type of information.
- Prepare the application packet in a format that can be painlessly updated (e.g. on a computer).
- Continue to update the packet as new information about your qualification, expertise and experience become available.
- Break the packet up into relevant sections with enlightening headings. For example, an academic employment packet may include section headings such as 'Research performance', 'Publications record', 'Service activities', and 'Teaching experience'.
- Make the information concise and easily discernible. Employers deal with hundreds of applications. Not many of them will have time to decipher complex and lengthy documents.

10.4 PERSONAL AND PROFESSIONAL REFERENCE LISTS

- Identify individuals whom you can trust to write great things about you. The key to finding such individuals is to have already proved your worthiness to them long before you need their references.
- Be careful about using casual acquittances, people you do not know well, and people who do not know you very well, as job referees. Even though they can write nice things (which are probably true) about you, experienced employers can easily tell when reference letters sound too glowing or superficial. Good

letters are those that come from the heart with detailed comments about your specific skills, based on the writer's direct knowledge of you.

- Do not 'brown-nose' your referees. Your good record is enough to earn you excellent comments from the referees.
- For job advertisements that welcome nominations, the best approach is for your thesis advisor to make the initial contacts for you. He or she may send the employer an abridged version of your resumé to give the employer a preview of your qualifications. You will then follow up with your formal application.
- Contact the potential referees ahead of time to seek their consent for providing references for you. Let them know the time frame over which you are likely to need their services.

10.5 LOCATING PROSPECTIVE EMPLOYERS

- Remember that the job market can be very volatile. The job opportunities available to you may vary from year to year as a function of the prevailing economic situation.
- Consult trade magazines in your field to find out what jobs are currently available.
- Attend relevant professional conferences and actively participate in discussion groups. Serve as a speaker whenever there is an opportunity. Attend open or 'invited' social cocktails at the conferences and network.
- Watch out for announcements sent directly to your department. Such announcements appear long before advertisements appear in publications.
- Use the computerized services that are becoming increasingly available. On-line services such as CompuServe can be useful for prompt identification of job sources.
- Network with friends and professionals to get inside information about job openings.
- Participate in job placement services of professional organizations. Many of these services are free of charge.
- Consult faculty members in your department about job openings they may be aware of.
- Be selective on the basis of where and which employer will make you happy. Remember that the job may turn out to

be a life-long commitment. Your long-term happiness is very essential. Seriously consider where you want to live and work.

- Be wary of job advertisements that have too imminent a closing date at the time of the appearance of the advertisement. It could mean that the employers have already identified 'their' candidate for the job and they are just advertising to boost the application pool or to comply with legal or procedural requirements.
- Do not apply for jobs for which you clearly do not have the background or requirements that the employer wants.
- Be enthusiastic about each job you apply for. Once you decide to apply, go all the way with the effort. A professional should never appear nonchalant or complacent in any of his/her endeavors.
- Send your application materials by certified or return receipt mail, if possible, so that you can get acknowledgement of receipt. If you cannot send it by certified mail and you do not receive acknowledgement from the employer within a reasonable time, contact them to ascertain receipt.

10.6 INTERVIEW PREPARATION

- Remember that the job market is always competitive. You have to beat out the competition. Every little detail counts.
- Use mock interviews, if necessary, to prepare for job interviews. For academic job search, classroom presentations can be used for this purpose.
- Always write thank you notes to the employer after your interview.
- Equip yourself for the interview by having the following:
  (a) copy of your application packet;
  (b) audio-visual aids (e.g. overhead slides) for presentations;
  (c) copies of your publications;
  (d) documentation about your teaching experience;
  (e) information about your student activities;
  (f) any materials the employer requests you to bring;
  (g) documentation about your legal working status, if needed.
- Prepare for typical interview questions such as:
  (a) How would you characterize yourself?
  (b) What do you know about this organization?

(c) Why do you want to work for us?

(d) What was your biggest accomplishment during your education?

(e) What are your strengths and weaknesses?

Questions for academic job seekers:

(a) In what journals do you expect to publish your work?

(b) Which sources of research funding do you plan to explore?

(c) Which professional organizations do you plan to belong to?

- Avoid giving canned or untruthful responses to interview questions.
- Ask probing questions such as the following:

(a) What are the employer's goal and primary objectives?

(b) What is the background composition of the company?

(c) What are the promotion requirements?

(d) What are the maternity and/or paternity leave policies?

(e) What family or group programs are available?

(f) What supporting services are available for the position?

(g) Does the company cover relocation expenses?

(h) What insurance program is available?

(i) What is the retirement package?

10.7 CONSIDERING A JOB OFFER

Do not let money alone dictate your decision. This is a difficult temptation to overcome for first-time job seekers. Given the human hierarchy of needs, money will cease to be the most important factor to a professional after a few years. Negotiate for the highest salary possible, but do not neglect your other present and future needs.

The job offer should be carefully evaluated as a package.

10.7.1 If the offer does not come

Be prepared for rejection. Not all job search efforts are successful on first trial. Remember that success is failure turned inside out. Take each rejection as an opportunity to intensify your job search effort. The harder you try, the better you will appear.

10.7.2 Reasons for changing jobs

- significant new responsibility offered by the new job;
- higher compensation;
- more productive use of training and experience;
- variety of experience offered by the new job;
- opportunity for advancement;
- new training opportunity;
- better professional environment;
- support for additional (higher) educational pursuits;
- better company benefits
- cost of living in new area;
- geographic preference;
- image of the new company.

10.8 TRANSFERRING RESEARCH TO PRACTICAL APPLICATION

Project and organizational transfers are important aspects of research. The organizational structure that is in effect during a project can influence the transfer of the final product of a project. The organic (internal) organizational structure and the external organizational structure must be linked by a discernible transfer path.

Research transfers may involve a complete project or components of a project. One possibility involves how products, ideas, concepts and decisions transfer from one project environment to another. The receiving organization (referred to as the transfer target) uses the transferred elements to generate new products, ideas, concepts, and decisions, which follow a reverse transfer path to the transfer source. Thereby, both project environments operate on a symbiotic basis, with each contributing something to the other.

Project transfer can be achieved in various forms. Three possible transfer modes that represent basic strategies for getting research result from one point to another are discussed below:

1. **Transfer of complete research results**: in this case, a fully developed research product is transferred from one project to another project. Very little product development effort is carried out at the receiving point. However, information about the operations of the product is fed back to the source so

that necessary product enhancements can be pursued. So the recipient generates product information which serves as a resource for future work of the transfer source.

2. **Transfer of research procedures and guidelines:** in this transfer mode, procedures and guidelines are transferred from one research environment to another. The project blueprints are implemented locally to generate the desired services and products. The use of local raw materials and manpower is encouraged for the local operations. Under this mode, the implementation of the transferred procedures can generate new operating procedures that can be fed back to enhance the original project. With this symbiotic arrangement, a loop system is created whereby both the transferring and the receiving organizations obtain useful benefits.

3. **Transfer of research concepts, theories, and ideas:** this strategy involves the transfer of the basic concepts, theories and ideas associated with a project. The transferred elements can then be enhanced, modified or customized within local constraints to generate new project outputs. The local modifications and enhancements have the potential to initiate an identical project, a new related project, or a new set of project concepts, theories, and ideas. These derived products may then be transferred back to the transfer source.

The important questions to ask about research transfer include the following:

- What exactly is being transferred?
- Who is receiving the transferred elements?
- What is the cost of the transfer?
- How is this research similar to previous projects?
- How is this research different from previous ones?
- Are the goals of the projects similar?
- What is expected from the transferred results?
- Is there adequate in-house skill to use the transferred results?
- Is the prevailing management culture receptive to the new project?
- Is the current infrastructure capable of supporting the project?
- What modifications will be necessary to the original research?

Sample technical paper

Reprinted with permission of John Wiley & Sons, Inc.
Badiru, Adedeji B., 'Multifactor Learning and Forgetting Models for Productivity and Performance Analysis,' *International Journal of Human Factors in Manufacturing*, Vol. 4, No. 1, 1994, pp. 37–54.

Multi-Factor Learning and Forgetting Models for Productivity and Performance Analysis

Adedeji B. Badiru
School of Industrial Engineering, University of Oklahoma, Norman, OK 73019, USA

ABSTRACT

This paper presents an approach to modeling multi-factor learning curve models for productivity and performance analysis in manufacturing. The models account for alternate periods of learning and forgetting and the resulting performance of an operator. Multi-factor learning curves facilitate the inclusion of more than one important factor in performance and productivity analysis. The inclusion of a component that accounts for the rate of forgetting establishes a realistic representation of the effects of learning and forgetting on performance. Analyses based on the proposed approach can facilitate more reliable management decisions in productivity analysis. Two-factor models are used to illustrate cases where forgetting occurs continuously or intermittently over a given time period.

Keywords: Learning curve, Experience curve, Performance analysis, Productivity analysis, Forgetting

INTRODUCTION

Learning is the essence of progress, forgetting is the root of regression.

Learning and forgetting are natural phenomena that directly affect productivity and performance. While extensive literature abounds on the subject of learning, no rigorous analytical study of the effect of forgetting can be found in the literature. Learning curve, also known as experience curve, represents the improved performance achieved through repeated performance of a specific function. Performance improvement due to the effect of learning has been extensively addressed in the literature over the past several decades. Examples include Wright (1936), Hoffman (1950), Preston and Keachie (1964), Hirchman (1964), Spradlin and Pierce (1967), Oi (1967), Conley (1970), Baloff (1971), Washburn (1972), Abernathy and Wayne (1974), Knecht (1974), Belkaoui (1976), Ebert (1976), Sule(1978), Richardson (1978), Imhoff (1978), Towill and Kaloo (1978), Nanda (1979), Liao (1979), Yelle (1979, 1980, 1983), Howell (1980), Fisk and Ballou (1982), Chen (1983), Kopcso and Nemitz (1983), Jewell (1984), Globerson and Shtub (1984), Belkaoui (1986), Smunt (1986), Camm, Evans, and Womer (1987), Womer and Gulledge (1983), Smith (1989), Dada and Srikanth (1990), and Badiru (1991).

A pioneering study by Wright (1936) confirmed the '80 percent learning effect' in airplane production plants. This indicates that a given operation is subject to a 20 percent productivity improvement whenever cumulative production doubles. Because of the several variables that interact in manufacturing systems (Badiru 1988), there is a need for multivariate learning curves. There has been much research interest in multivariate progress functions (Conway and Schultz 1959; Alchian 1963; Preston and Keachie 1964; Glover 1966; Graver and Boren 1967; Goldberger 1968; Carlson 1973; McIntyre 1977; Womer 1979; Cox and Gansler 1981; Bemis 1981; Womer 1981; Donath *et al.* 1981; Gold 1981; Waller and Dwyer 1981; Gulledge, Womer, and Dorroh 1984; Gulledge *et al.* 1985; Gulledge and Khoshnevis 1987; Camm, Gulledge, and Womer 1987; Dada and Srikanth 1990). Badiru (1991) presents an application of multivariate learning curves to manufacturing cost estimation.

A major issue that has not been adequately addressed in previous works is the fact that workers also forget during the process of learning. Sule (1978) presents one of the earliest documented studies of alternate periods of learning and forgetting using a univariate learning curve. This paper extends the univariate model to multivariate models. The proposed approach accounts for alternate periods of learning and forgetting. Multivariate analysis facilitates the inclusion of additional important factors in manufacturing productivity analysis. The inclusion of a forgetting model creates a realistic representation of manufacturing operations that are subject to interruption in the learning process. Learning tends to improve performance while forgetting tends to reduce performance.

Multivariate learning curve analysis facilitates the inclusion of additional important factors in productivity analysis. The consideration of a forgetting model creates a realistic representation of operations that are subject to interruptions in the learning process (Smith, 1989). This is contrary to the conclusion by Womer (1984) that production breaks do not necessarily induce a loss of learning. In this paper, it is concluded that production interruptions can, indeed, cause a loss in learning if during the inter- ruptions workers are involved in non-complementary operations, the learning process of which counteracts the preceding learning process. For example, shifting from learning one computer programming syntax to another can lead to loss of learning. Important research questions that should be addressed include the following:

- What factors influence learning?
- What factors influence forgetting?
- What joint effects do learning and forgetting have on worker perfor- mance and productivity?

In this paper, a brief review of univariate and multivariate learning curve models is first presented. A bivariate model is used to illustrate the general nature of multivariate learning curve models. The introduction of a forgetting component into the learning phenomenon is then discussed. The conflicting tendencies of learning and forgetting phenomena are modeled by a functional resolution approach. Bivariate models are used to illustrate cases where forgetting occurs continuously or intermittently over a specified time period. In the last portion of the paper, a numerical example is presented to illustrate the effect of forgetting on the overall productivity of a worker.

UNIVARIATE AND MULTIVARIATE LEARNING CURVES

The univariate learning curve expresses a dependent variable (e.g. pro- duction cost) in terms of some independent variable (e.g. cumulative production). Numerous learning curve models have been presented in the literature. The most notable univariate models include the log-linear model (Wright 1936), the S-curve (Carr 1946), the Stanford-B model (Asher 1956), DeJong's learning formula (DeJong 1957), Levy's ad- aptation function (Levy 1965), Glover's learning formula (Glover 1966), Pegels' exponential function (Pegels 1969, 1976), Knecht's upturn model (Knecht 1974), and Yelle's combined product model (Yelle 1976). The log-linear model is recognized as the basic model for most manufacturing productivity analysis. The model presents the relationship between cumulative average cost per unit and cumulative production as shown below:

$$C_x = c_1 x^b \qquad (1)$$

$$\log C_x = \log c_1 + b \log x \qquad (2)$$

where C_x = cumulative average cost of producing x units, c_1 = cost of the first unit, x = cumulative production, and b = the learning curve exponent. The expression for the learning rate, p, is defined as the percent productivity gain based on two production levels, x_1 and x_2, where $x_2 = 2x_1$. It is computed as C_{x2}/C_{x1}, which yields the relationship below:

$$p = 2^b \tag{3}$$

Extensions of the univariate learning curve are important for realistic analysis of productivity gain. In manufacturing operations, several quantitative and qualitative factors intermingle to compound performance analysis. Heuristic decision making, in particular, requires a careful consideration of qualitative factors. There are numerous factors that can influence how fast, how far and how well a worker learns within a given time span. Multivariate models have a significant place in manufacturing performance analysis. One form of multivariate learning curve (Badiru 1992) is defined by:

$$C_x = K \prod_{i=1}^{n} c_i x_i^{b_i} \tag{4}$$

where C_x = cumulative average cost per unit, K = cost of first unit of the product, x = vector of specific values of independent variables, x_i = specific value of the ith factor, n = number of factors in the model, c_i = coefficient for the ith factor, and b_i = learning exponent for the ith factor. A bivariate form of the model is presented below:

$$C = \beta_0 x_1^{\beta_1} x_2^{\beta_2} \tag{5}$$

where C is a measure of cost and x_1 and x_2 are independent variables of interest. Conway and Schultz (1959) first suggested the need for a multivariate generalized learning curve. Some of the multivariate models that have been reported in the literature are discussed briefly below. Alchian (1963) modeled learning curves that estimate direct labor per pound of airframe needed to manufacture the Nth airframe in a cumulative production of N planes based on World War II data. He studied the alternate functions presented below to describe the relationships between direct labor per pound of airframe (m), cumulative production (N), time (T), and rate of production per month (ΔN):

$$
\begin{aligned}
\log m &= a_2 + b_2 T \\
\log m &= a_3 + b_3 T + b_4 \Delta N \\
\log m &= a_4 + b_5 (\log T) + b_6 (\log \Delta N) \\
\log m &= a_5 + b_7 T + b_8 (\log \Delta N) \\
\log m &= a_6 + b_9 T + b_{10} (\log N) \\
\log m &= a_7 + b_{11} (\log N) + b_{12} (\log \Delta N)
\end{aligned}
\tag{6}
$$

The multiplicative power function, often referred to as the Cobb – Douglas function, was investigated by Goldberger (1968) as a model for a learning curve. The model is of the general form below:

$$C = b_0 x_1^{b_1} x_2^{b_2} \ldots x_n^{b_n} \varepsilon \tag{7}$$

where C = estimated cost, b_0 = model coefficient, x_i = ith independent variable $(i = 1, 2, \ldots, n)$, b_i = exponent of the ith variable, and ε = error component in the model. For parametric cost analysis, Waller and Dwyer (1981) present an additive model of the form:

$$C = c_1 x_1^{b_1} + c_2 x_2^{b_2} + \ldots + c_n x_n^{b_n} + \varepsilon \tag{8}$$

where c_i $(i = 1, 2, \ldots, n)$ is the coefficient of the ith independent variable. The model was reported to have been fitted successfully for missile tooling and equipment cost. A variation of the power model was used by Bemis (1981) to study weapon system production. Cox and Gansler (1981) discuss the use of a bivariate model for the assessment of the costs and benefits of a single-source versus multiple-source production decision with variations in quantity and production rate in major DOD (Department of Defense) programs. A similar study by Camm, Gulledge, and Womer (1987) also uses the multiplicative power model to express program costs in terms of cumulative quantity and production rate in order to evaluate contractor behavior.

McIntyre (1977) introduced a nonlinear cost-volume-profit model for learning curve analysis. The nonlinearity in the model is effected by incorporating a nonlinear cost function that expresses the effects of employee learning. The profit equation for the initial period of production for a product subject to the usual learning function is expressed as:

$$P = px - c(ax^{b+1}) - f \tag{9}$$

where P = profit, p = price per unit, x = cumulative production, c = labor cost per unit time, f = fixed cost per period, and b = index of learning. The profit function for the initial period of production with n production processes operating simultaneously is given as:

$$P = px - nca\left(\frac{x}{n}\right)^{b+1} - f \tag{10}$$

where x is the number of units produced by n labor teams consisting of one or more employees each. Each team is assumed to produce x/n units. This model indicates that when additional production teams are included, more units are produced over a given time period. However, the average time for a given number of units increases because more employees are producing while they are still learning. That is, more employees with low (but improving) productivity are engaged in production at the same time. The preceding model is extended to the case where employees with different skill levels produce different learning parameters between production runs. This is modeled as:

$$P = p \sum_{i=1}^{n} x_i - c \sum_{i=1}^{n} a_i x_i^{b_i+1} - f \tag{11}$$

where a_i and b_i denote the parameters applicable to the average skill level of the ith production run and x_i represents the output of the ith run in a given time period. This model could be useful for manufacturing systems that call for concurrent engineering.

Womer (1979) presents a multivariate model that incorporates cumulative production, production rate, and program cost. His approach involves a production function that relates output rate to a set of inputs with variable utilization rates as presented below:

$$q(t) = AQ^\delta(t)x^{1/\gamma}(t) \tag{12}$$

where, A = constant, $q(t)$ = program output rate at time t, $Q(T)$ = cumulative production at time T, δ = learning parameter, γ = returns to scale parameter, and $x(t)$ = rate of variable resource utilization at time t. To optimize the discounted program cost, the cost function is defined as:

$$C = \int_0^T x(t)e^{-pt}dt, \tag{13}$$

where p is the discount rate and T is the time horizon for the analysis. If V is defined as the planned cumulative production at time T (i.e. $Q(T) = V$), then the problem can be formulated as the following optimization problem:

$$\min \int_0^T x(t)e^{-pt}dt \tag{14}$$

subject to:

$$\begin{aligned} q(t) &= AQ^\delta(t)x^{1/\gamma}(t) \\ x(t) &\geq 0 \\ Q(0) &= 0 \\ Q(T) &= V \end{aligned}$$

whose solution yields the estimated cost at time t, given V and T:

$$C(t|V,T) = [p/(\gamma - 1)]^{\gamma-1}(1 - \delta)^{-\gamma}A^{-\gamma}V^{\gamma(1-\delta)}[e^{pT/(\gamma-1)} - 1]^{-\gamma}[e^{pt/(\gamma-1)} - 1] \tag{15}$$

Example of a bivariate model

A bivariate model adapted from Badiru (1991) is used in this section to illustrate the nature and modeling approach for general multivariate models. A model presented by Badiru (1991) involves a learning curve containing two independent variables: cumulative production (x_1) and cumulative training time (x_2). The following hypothesized model was chosen for illustrative purposes:

$$C_{x_1 x_2} = K c_1 x_1^{b_1} c_2 x_2^{b_2} \tag{16}$$

where C = cumulative average cost per unit for a given set of factors, K = intrinsic constant, x_i = specific value of the ith factor, c_i = coefficient for the ith factor, and b_i = learning exponent for the ith factor. The set of test data used for the modeling is shown in Table 1. Two data replicates are used for each of the ten combinations of cost and time values. Observations are recorded for the number of units representing double production volumes. The model is transformed to the logarithmic form below:

$$\log C_x = \log a + b_1 \log x_1 + b_2 \log x_2 \tag{17}$$

where $a = (K)(c_1)(c_2)$. A regression approach yielded the fitted model below:

$$\log C_X = 5.70 - 0.21(\log x_1) - 0.13(\log x_2) \tag{18}$$

That is:

$$C_X = 298.88 x_1^{-0.21} x_2^{-0.13} \tag{19}$$

where $\log(a) = 5.70$ (i.e. $a = 298.88$), x_1 = cumulative production units, and x_2 = cumulative training time in hours. Figure 1 shows the response surface for the fitted model. As in the univariate case, the bivariate model indicates that the cumulative average cost decreases as cumulative pro-

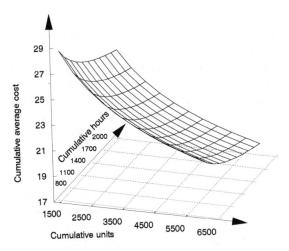

Figure 1 Effect of production volume and training hours on performance.

Table 1 Data for modeling bivariate learning curve

Treatment number	Observation number	Cumulative average cost ($)	Cumulative production (Units)	Cumulative training time (Hours)
1	1	120	10	11
	2	140	10	8
2	3	95	20	54
	4	125	20	25
3	5	80	40	100
	6	75	40	80
4	7	65	80	220
	8	50	80	150
5	9	55	160	410
	10	40	160	500
6	11	40	320	660
	12	38	320	600
7	13	32	640	810
	14	36	640	750
8	15	25	1280	890
	16	25	1280	800
9	17	20	2560	990
	18	24	2560	900
10	19	19	5120	1155
	20	25	5120	1000

duction and training time increase. The 95% confidence intervals for the parameters in the model are shown in Table 2.

The result of analysis of variance for the full regression is presented in Table 3. The P-value of 0.0000 in the table indicates that we have a highly significant regression fit. The R-squared value of 0.966962 indicates that most of the variabilities in cumulative average cost are explained by the terms in the model. Table 4 shows the breakdown of the model component of the sum of squares. Based on the low P-values shown in the table, it is concluded that both production quantity and training time contribute significantly to the regression model.

The correlation matrix for the estimates of the coefficients in the model is shown in Table 5. It is seen that log of units and log of time are very negatively correlated and the constant is positively correlated with log of units while it is negatively correlated with log of time. The strong negative correlation (-0.9189) between units and training time suggests that there

Table 2 95% confidence interval for model parameters

Parameter	Estimate	Lower limit	Upper limit
$\log(a)$	5.7024	5.4717	5.9331
b_1	−0.2093	−0.2826	−0.1359
b_2	−0.1321	−0.2269	−0.0373

Table 3 ANOVA table for the fitted model

Source	Sum of squares	df	Mean square	F-ratio	P-value
Model	7.41394	2	3.70697	248.778	0.0000
Error	0.253312	17	0.00149007		
Total	7.66725	19			

R-squared = 0.966962, R-adjusted (adjusted for degrees of freedom) = 0.963075, Standard error of estimate = 0.122069

Table 4 ANOVA breakdown of model components

Source	Sum of squares	df	Mean square	F-ratio	P-value
Log (units)	7.28516187	1	7.2851619	488.91	0.000
Log (time)	0.12877864	1	0.1287786	8.64	0.0092
Model	7.41394052	2			

Table 5 Correlation matrix for coefficient estimates

	Constant	Log (units)	Log (time)
Constant	1.0000	0.3654	−0.6895
Log (units)	0.3654	1.0000	−0.9189
Log (time)	−0.6895	−0.9189	1.0000

is strong multicollinearity. Multicollinearity normally implies that one of the correlated variables can be omitted without jeopardizing the fit of the model. Variables that are statistically independent will have an expected correlation of zero. As expected, Table 5 does not indicate any zero correlations. The source of strong correlations may be explained by the fact that it is difficult to separate the effects of training time from the effect of cumulative production. Obviously, the level of training will influence productivity, which may be reflected in the level of cumulative production within a given length of time.

INCORPORATION OF FORGETTING FUNCTIONS

Retention rate and retention capacity of different workers will determine the nature of the forgetting function to be modeled for the workers. Whenever interruption occurs in the learning process, it results in some forgetting. The resulting drop in performance rate depends on the initial level of performance and the length of the interruption. There are three potential cases for the occurrence of forgetting:

1. Forgetting occurs continuously throughout the learning process.
2. Forgetting occurs only over a bounded time interval.
3. Forgetting occurs over intermittent time intervals where the time of occurrence and the duration of forgetting are described by some probability distributions.

The first two cases are illustrated in this paper. The third case is a matter for further research investigations. First, the univariate case is used to illustrate the concept of learning and forgetting. Then, the bivariate case is used to illustrate the modeling and computational analysis procedures. The illustrative examples can be extended to general multivariate models.

The univariate case

Any operation that is subject to interruption in the learning process is suitable for the application of forgetting functions. Sule (1978) postulated that the forgetting model can be represented as:

$$Y_f = X_f R_f^{B_f} \tag{20}$$

where Y_f = number of units that could be produced on Rth day, X_f = equivalent production on first day of the forgetting curve, R_f = cumulative number of days in forgetting cycle, and B_f = forgetting rate. This forgetting model is of the same form as the standard progress function except that the forgetting rate will be negative where the learning rate is positive and vice versa. Possible forms of univariate forgetting functions are shown in Figure 2. Model (a) shows a case where the worker forgets rapidly from an initial performance level. Model (b) shows a case where forgetting occurs more slowly in a concave fashion. Model (c) shows a case where there is some residual retention of performance after a period of progressive forgetting.

The combination of the learning and forgetting models will present a more realistic picture of what actually occurs in a learning process. The combination is not as simple as resolving two curves to obtain a resultant curve. The resolution is particularly complex in the case of intermittent periods of forgetting. Figure 3 presents a view of some periods where

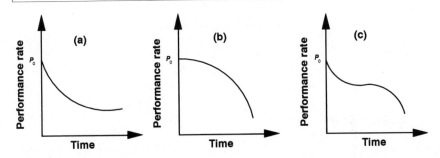

Figure 2 Univariate models of forgetting.

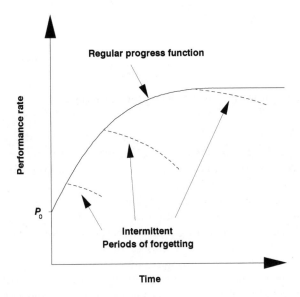

Figure 3 Intermittent period of forgetting.

forgetting takes place. Figure 4 presents what the resultant learning/forgetting curve might look like.

The bivariate case

Hypothetical examples of bivariate learning function, $l(t,u)$, and forgetting function, $f(t,u)$, are shown respectively below:

$$l(t, u) = 20t^{0.09} + u^{-0.05} \tag{21}$$

$$f(t, u) = t^{-0.20}u^{-0.30} \tag{22}$$

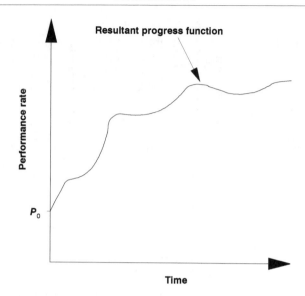

Figure 4 Resultant learning/forgetting performance curve.

for $100 \leq t \leq 3000$ and $100 \leq u \leq 1000$, where t represents time and u represents production units. Figure 5 shows the learning function while Figure 6 shows the forgetting function. In multivariate cases, we will be referring to performance surfaces rather than performance curves. The positive effect of learning is represented in terms of performance rate. Thus, in Figure 5, performance rate increases with time. The forgetting function may be viewed as the model of what the performance level would be if no additional learning occurs. Thus, if learning terminates at a particular performance level, the effect of forgetting will gradually reduce that initial performance level based on the functional form of the forgetting model. Due to the effect of forgetting, the performance level tends to decrease with time. An approach used to resolve the surfaces is presented in the following sections.

Case of continuous forgetting

If learning and forgetting start at a particular performance level, the resultant performance function, $r(t,u)$, may be modeled as:

$$r(t, u) = l(t, u) - \frac{[l(t, u) - f(t, u)]}{2}$$

$$= \frac{l(t, u) + f(t, u)}{2} \tag{23}$$

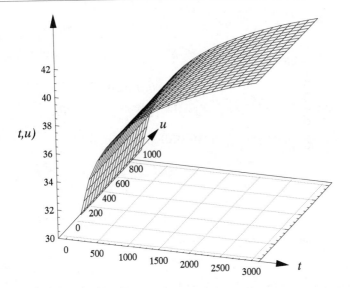

Figure 5 Bivariate model of increasing performance due to learning.

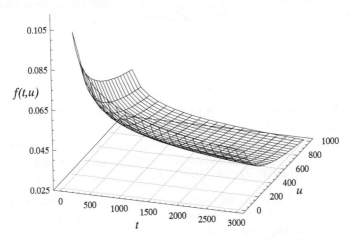

Figure 6 Bivariate model of decreasing performance due to forgetting.

This is simply the point-by-point average of the learning and forgetting functions. The justification for using the above approach for resolving the two functions can be seen by considering the univariate curves in Figure 7. In the figure, the learn function, $l(t)$, is above the forgetting function, $f(t)$. The forgetting function will create a downward pull on the learn

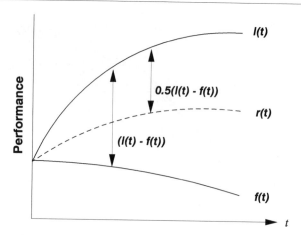

Figure 7 Learning/forgetting curve resolution approach.

function. This creates the resultant function, $r(t)$. Note that the resolution of the two functions is applicable only for the time periods over which forgetting actually occurs.

Using the above resolution approach for the illustrative examples of $l(t,u)$ and $f(t,u)$, we obtain the following resultant function for the case of continuous forgetting:

$$r(t, u) = 20t^{0.09} + u^{-0.05} - \frac{[20t^{0.09} + u^{-0.05} - t^{-0.20}u^{-0.30}]}{2}$$

$$= 10t^{0.09} + 0.5u^{-0.05} + 0.5t^{-0.2}u^{-0.3} \qquad (24)$$

It should be noted that $l(t,u)$ and $f(t,u)$ do not really start at the same performance level since the expected performance at $t = 0$ for $l(t,u)$ is much higher than the expected performance at $t = 0$ for $f(t,u)$. Thus, alternate resolution approaches need to be investigated for this type of function. This leads us to an important definition:

Coincident learning and forgetting functions: a learning function and a forgetting function are said to be coincident if both functions originate at the same performance level.

For noncoincident functions, alternate resolution approaches may be considered whereby the starting point of the forgetting function is incorporated into the resolution procedures. For example, an integer multiple of the learning function may be used (relative to the forgetting function) during the resolution process. In a specific case, the learning function may be given N times as much weight as the forgetting function in obtaining the point-by-point average of the functions. Another approach is to define the forgetting function as a relative function based directly on the

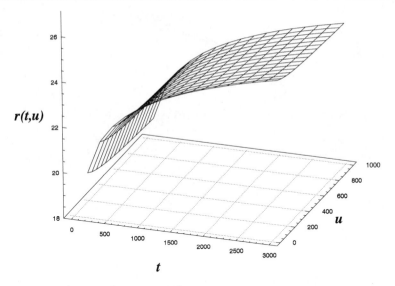

Figure 8 Resultant performance surface.

level of decrement it creates in the overall performance. In this case, the two functions may be added directly to obtain the resultant performance function.

A plot of the resultant performance surface, $r(t,u)$, for the case of continuous forgetting is presented in Figure 8, which shows that the effect of forgetting has reduced the resultant performance levels. It is noted that the forgetting function in the above example is defined as an absolute function that starts at low performance levels. The low starting levels result in net performance levels that are about half the original performance level without the effect of forgetting.

Bounded interval of forgetting

Suppose forgetting only occurs over the bounded time interval $500 \leq t \leq 1500$ in the preceding example. Figure 9 shows the bounded portion of the learning function, $l(t,u)$, while Figure 10 shows a bounded portion of the forgetting function, $f(t,u)$. This bounded time interval represents a region of steep performance improvement. Figure 11 shows the resultant performance function over the bounded interval.

Case of intermittent forgetting

In this case, forgetting occurs in an intermittent fashion. Each occurrence is described by the prevailing forgetting function. It is also possible to

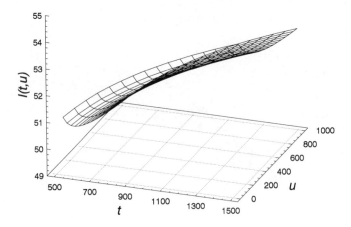

Figure 9 Learning function over $500 \leq t \leq 1500$.

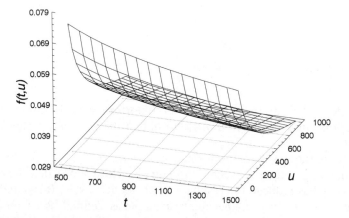

Figure 10 Forgetting function over $500 \leq t \leq 1500$.

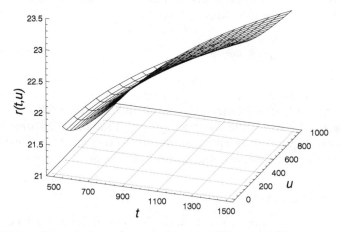

Figure 11 Resultant performance function over $500 \leq t \leq 1500$.

have alternate periods of forgeting and forgetting. The plot of the resultant bivariate model for this case will exhibit a 'wave-like' form (refer to Figure 4 for the univariate case). This particular case is not within the scope of the present study.

Numerical example

It was noted earlier that the functions used in the illustrative example were not coincident. Coincidence can be achieved by shifting up the forgetting function by a scalar quantity such that its highest point coincides with the lowest point of the learning function. This common point will be referred to as the incipient performance level, P_i. For our example, this approach yields a shifted function, $f_s(t,u)$, and an adjusted resultant function, $r_a(t,u)$, defined as follows:

$$f_s(t, u) = 31.0 + t^{-0.20}u^{-0.30} \tag{25}$$

$$r_a(t, u) = 15.5 + 10t^{0.09} + 0.5u^{-0.05} + 0.5t^{-0.2}u^{-0.3} \tag{26}$$

for $100 \leqslant t \leqslant 3000$ and $100 \leqslant u \leqslant 1000$. Figure 12 shows the learning function, the shifted forgetting function, and the adjusted resultant function on the same plot. Though not drawn to scale, the plot shows the relative positions and shapes of the three functions. The resultant function can be used for various practical applications. An example is the evaluation of production standards as illustrated below. Let the units of performance expressed by $r_a(t,u)$ be in terms of hundreds of assembly components produced per cycle while u represents cumulative units of completed assemblies and let t represent cumulative hours of operation. Table 6 presents a sample of the tabulation of values of $r_a(t,u)$ for various combinations of values of t and u.

Application problem: given that the standards department of a manufacturing plant has set a target production rate of 4000 (i.e. $r = 40.00$) components per cycle to be achieved after 2000 hours (i.e. $t = 2000$) of operation and 670 cumulative production units (i.e. $u = 670$). Determine feasibility of the target performance. Using the fitted model of $r_a(t,u)$, we have:

$$r(2000, 670) = 15.5 + 10(2000)^{0.09} + 0.5(670)^{-0.05} + 0.5 (2000)^{-0.2}(670)^{-0.3}$$

which yields a performance rate of 3570 components per cycle as seen in Table 6. Based on this result, it can be determined that the target performance is not feasible. Note that without the effect of forgetting, the performance rate is computed to be:

$$l(2000, 670) = 20(2000)^{0.09} + (670)^{-0.05}$$
$$= 40.36$$

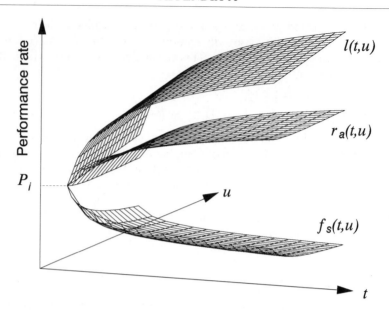

Figure 12 Resolution of coincident learning and forgetting functions.

Table 6 Tabulated values of $r_a(t, u)$

t	u	$r_a(t,u)$	t	u	$r_a(t,u)$
100	100	3108	1600	550	3531
200	130	3204	1700	580	3541
300	160	3263	1800	610	3551
400	190	3306	1900	640	3561
500	220	3341	2000	670	3570
600	250	3369	2100	700	3578
700	280	3393	2200	730	3586
800	310	3415	2300	760	3594
900	340	3434	2400	790	3602
1000	370	3451	2500	820	3609
1100	400	3467	2600	850	3616
1200	430	3482	2700	880	3623
1300	460	3495	2800	910	3630
1400	490	3508	2900	940	3636
1500	520	3520	3000	970	3642

That is, 4036 components per production cycle, which exceeds the target performance level. Thus, under regular learning without the effect of forgetting, it is possible to achieve higher performance levels. With the detrimental effect of forgetting, actual performance may not be as high as we might normally expect. This hypothetical example illustrates the importance of considering the effect of forgetting when evaluating performance on the basis of learning curves. In any practical situation, an allowance must be made for the potential impacts that forgetting may have on performance. Potential applications of the combined learning and forgetting models include design of training programs, manufacturing economic analysis, manpower scheduling, production planning, labor estimation, budgeting, and resource allocation.

CONCLUSION

This paper provides a new insight into the effect of the interaction of learning and forgetting on worker performance. The paper presents an approach to developing a bivariate model of manufacturing progress function that accounts for alternate periods of learning and forgetting. The approach incorporates a bivariate forgetting model into a bivariate learning model. Bivariate analysis facilitates the inclusion of additional important factors in performance and productivity analysis. The inclusion of a forgetting model facilitates a realistic representation of production operations that are subject to interruption (forgetting) during the learning processes. Additional research is needed to study the mathematical complexity of combining multivariate learning models and intermittent multivariate forgetting models.

REFERENCES

Abernathy, W. J. and Wayne, K. 'Limits of the Learning Curve,' *Harvard Business Review*, Vol. 52, Sept-Oct, 1974, pp. 109–119.

Alchian, Armen, 'Reliability of Progress Curves in Airframe Production,' *Econometrica*, Vol. 31, No. 4, 1963, pp. 679–693.

Asher, Harold, *Cost–Quantity Relationships in the Airframe Industry*, Report No. R-291, The Rand Corporation, Santa Monica, CA, July 1, 1956.

Badiru, Adedeji B., *Project Management in Manufacturing and High Technology Operations*, John Wiley & Sons, New York, 1988.

Badiru, Adedeji B., 'Manufacturing Cost Estimation: A Multivariate Learning Curve Approach,' *Journal of Manufacturing Systems*, Vol. 10, No. 6, 1991, pp. 431–441.

Badiru, Adedeji B., 'Computational Survey of Univariate and Multivariate Learning Curve Models,' *IEEE Transactions on Engineering Management*, Vol. 39, No. 2, May 1992, pp. 176–188.

Baloff, Nicholas, 'Extension of the Learning Curve: Some Empirical Results,' *Operations Research Quarterly*, Vol. 22, No. 4, 1971, pp. 329–340.

Belkaoui, Ahmed, 'Costing Through Learning,' *Cost and Management*, Vol. 50, No. 3, 1976, pp. 36–40.

Belkaoui, Ahmed, *The Learning Curve*, Quorum Books, Westport, Conn., 1986.

Bemis, John C., 'A Model for Examining the Cost Implications of Production Rate,' *Concepts: The Journal of Defense Systems Acquisition Management*, Vol. 4, No. 2, 1981, pp. 84–94.

Camm, Jeffrey D., Evans, James R., and Womer, Norman K., 'The Unit Learning Curve Approximation of Total Cost,' *Computers and Industrial Engineering*, Vol. 12, No. 3, 1987, pp. 205–213.

Camm, Jeffrey D., Gulledge, Thomas R. Jr., and Womer, Norman K., 'Production Rate and Contractor Behavior,' *The Journal of Cost Analysis*, Vol. 5, No. 1, 1987, pp. 27–38.

Carlson, J. G. H., 'Cubic Learning Curves: Precision Tool for Labor Estimating,' *Manufacturing Engineering and Management*, Vol. 71, No. 5, 1973, pp. 22–25.

Carlson, J. G. H. and Rowe, A. J., 'How Much Does Forgetting Cost?,' *Industrial Engineering*, Vol. 8, September 1976, pp. 40–47.

Carr, G. W., 'Peacetime Cost Estimating Requires New Learning Curves,' *Aviation*, Vol. 45, April 1946.

Chen, J. T., 'Modeling Learning Curve and Learning Complementarity for Resource Allocation and Production Scheduling,' *Decision Sciences*, Vol. 14, 1983, pp. 170–186.

Conley, P., 'Experience Curves as a Planning Tool,' *IEEE Spectrum*, Vol. 7, No. 6, 1970, pp. 63–68.

Conway, R. W. and Schultz, Andrew Jr., 'The Manufacturing Progress Function,' *Journal of Industrial Engineering*, Vol. 1, 1959, pp. 39–53.

Cox, Larry W. and Gansler, J. S., 'Evaluating the Impact of Quantity, Rate, and Competition,' *Concepts: The Journal of Defense Systems Acquisition Management*, Vol. 4, No. 4, 1981, pp. 29–53.

Dada, Maqbool and Srikanth, K. N., 'Monopolistic Pricing and The Learning Curve: An Algorithmic Approach,' *Operations Research*, Vol. 38, No. 4, 1990, pp. 656–666.

DeJong, J. R., 'The Effects of Increasing Skill on Cycle Time and Its Consequences for Time Standards,' *Ergonomics*, November 1957, pp. 51–60.

Donath, N., Globerson, S. and Zang, I., 'A Learning Curve Model for Multiple Batch Production Process,' *International Journal of Production Research*, Vol. 19, No. 2, 1981, pp. 165–175.

Ebert, R. J., 'Aggregate Planning With Learning Curve Productivity,' *Management Science*, Vol. 23, 1976, pp. 171–182.

Fisk, J. C. and Ballou, D. P., 'Production Lot Sizing Under a Learning Effect,' *IIE Transactions*, Vol. 14, No. 4, 1982, pp. 257–264.

Globerson, S. and Shtub, A., 'The Impact of Learning Curves on the Design of Long Cycle Time Lines,' *Industrial Management*, Vol. 26, No. 3, May/June, 1984, pp. 5–10.

Glover, J. H., 'Manufacturing Progress Functions: An Alternative Model and Its Comparison with Existing Functions,' *International Journal of Production Research*, Vol. 4, No. 4, 1966, pp. 279–300.

Gold, B. 'Changing Perspectives on Size, Scale, and Returns,' *Journal of Economic Literature*, Vol. 19, No. 1, 1981, pp. 5–33.

Goldberger, Arthur S., 'The Interpretation and Estimation of Cobb-Douglas Functions,' *Econometrica*, Vol. 35, No. 3–4, 1968, pp. 464–472.

Graver, C. A. and Boren, H. E. Jr., *Multivariate Logarithmic and Exponential Regression Models,* RM-4879-PR, The Rand Corporation, Santa Monica, CA, 1967.

Gulledge, T. R. and Khoshnevis, B., 'Production Rate, Learning, and Program Costs: Survey and Bibliography,' *Engineering Costs and Production Economics*, Vol. 11, 1987, pp. 223–236.

Gulledge, Thomas R. Jr., Womer, Norman K., and Dorroh, J. R., 'Learning and Costs in Airframe Production: A Multiple Output Production Function Approach,' *Naval Research Logistics Quarterly*, Vol. 31, 1984, pp. 67–85.

Gulledge, Thomas R. Jr., Womer, Norman K., and Tarimcilar, M. Murat, 'A Discrete Dynamic Optimization Model for Made-to-Order Cost Analysis,' *Decision Sciences*, Vol. 16, 1985, pp. 73–90.

Hirchmann, W. B., 'Learning Curve,' *Chemical Engineering*, Vol. 71, No. 7, 1964, pp. 95–100.

Hoffman, F. S., *Comments on the Modified Form of the Air Craft Progress Functions*, Report No. RN-464, The Rand Corporation, Santa Monica, CA, 1950.

Howell, Sydney D., 'Learning Curves for New Products,' *Industrial Marketing Management*, Vol. 9, No. 2, 1980, pp. 97–99.

Imhoff, E. A. Jr., 'The Learning Curve and Its Applications,' *Management Accounting*, Vol. 59, No. 8, 1978, pp. 44–46.

Jewell, William S., 'A Generalized Framework for Learning Curve Reliability Growth Models,' *Operations Research*, Vol. 32, No. 3, May-June, 1984, pp. 547–558.

Knecht, G. R., 'Costing, Technological Growth and Generalized Learning Curves,' *Operations Research Quarterly*, Vol. 25, No. 3, Sept 1974, pp. 487–491.

Kopcso, David P., and Nemitz, William C., 'Learning Curves and Lot Sizing for Independent and Dependent Demand,' *Journal of Operations Management*, Vol. 4, No. 1, Nov. 1983, pp. 73–83.

Levy, F. K., 'Adaptation in the Production Process,' *Management Science*, Vol. 11, No. 6, April 1965, pp. B136–B154.

Liao, W. M., 'Effects of Learning on Resource Allocation Decisions,' *Decision Sciences*, Vol. 10, 1979, pp. 116–125.

McIntyre, E. V., 'Cost-Volume-Profit Analysis Adjusted for Learning,' *Management Science*, Vol. 24, No. 2, 1977, pp. 149–160.

Nanda, Ravinder, 'Using Learning Curves in Integration of Production Resources,' *Proceedings of 1979 IIE Fall Conference*, 1979, pp. 376–380.

Oi, W. Y., 'The Neoclassical Foundations of Progress Functions,' *Economic Journal*, Vol. 77, 1967, pp. 579–594.

Pegels, Carl C., 'On Startup or Learning Curves: An Expanded View,' *AIIE Transactions*, Vol. 1, No. 3, September 1969, pp. 216–222.

Pegels, Carl C., 'Start Up or Learning Curves – Some New Approaches,' *Decision Sciences*, Vol. 7, No. 4, Oct. 1976, pp. 705–713.

Preston, L. E. and Keachie, E. C., 'Cost Functions and Progress Functions: An Integration,' *American Economic Review*, Vol. 54, 1964, pp. 100–106.

Richardson, Wallace J., 'Use of Learning Curves to Set Goals and Monitor Progress in Cost Reduction Programs,' *Proceedings of 1978 IIE Spring Conference*, 1978, pp. 235–239.

Smith, Jason, *Learning Curve for Cost Control*, Industrial Engineering & Management Press, Norcross, GA, 1989.

Smunt, Timothy L., 'A Comparison of Learning Curve Analysis and Moving Average Ratio Analysis for Detailed Operational Planning,' *Decision Sciences*, Vol. 17, No. 4, 1986, pp. 475–495.

Spradlin, B. C. and Pierce, D. A., 'Production Scheduling Under a Learning Effect by Dynamic Programming,' *Journal of Industrial Engineering*, Vol. 18, No. 3, 1967, pp. 219–222.

Sule, D. R., 'The Effect of Alternate Periods of Learning and Forgetting on Economic Manufacturing Quantity,' *AIIE Transactions*, Vol. 10, No. 3, 1978, pp. 338–343.

Towill, D. R. and Kaloo, U., 'Productivity Drift in Extended Learning Curves,' *Omega*, Vol. 6, No. 4, 1978, pp. 295–304.

Waller, Eugene W., and Dwyer, Thomas J., 'Alternative Techniques for Use in Parametric Cost Analysis,' *Concepts: Journal of Defense Systems Acquisition Management*, Vol. 4, No. 2, Spring 1981, pp. 48–59.

Washburn, A. R., 'The Effects of Discounting Profits in the Presence of Learning in the Optimization of Production Rates,' *AIIE Transactions*, Vol. 4, 1972, pp. 205–213.

Womer, N. K., 'Estimating Learning Curves From Aggregate Monthly Data,' *Management Science*, Vol. 30, No. 8, 1984, pp. 982–992.

Womer, Norman K., 'Learning Curves, Production Rate, and Program Costs,' *Management Science*, Vol. 25, No. 4, 1979, pp. 312–319.

Womer, Norman K., 'Some Propositions on Cost Functions,' *Southern Economic Journal*, Vol. 47, 1981, pp. 1111–1119.

Womer, Norman K. and Gulledge, T. R. Jr., 'A Dynamic Cost Function for An Airframe Production Program,' *Engineering Costs and Production Economics*, Vol. 7, 1983, pp. 213–227.

Wright, T. P., 'Factors Affecting the Cost of Airplanes,' *Journal of Aeronautical Science*, Vol. 3, No. 2, February 1936, pp. 122–128.

Yelle, Louis E., 'Estimating Learning Curves for Potential Products,' *Industrial Marketing Management*, Vol. 5, No. 2/3, June 1976, pp. 147–154.

Yelle, Louis E., 'The Learning Curve: Historical Review and Comprehensive Survey,' *Decision Sciences*, Vol. 10, No. 2, April 1979, pp. 302–328.

Yelle, Louis E., 'Industrial Life Cycles and Learning Curves: Interaction of Marketing and Production,' *Industrial Marketing Management*, Vol. 9, No. 4, Oct 1980, pp. 311–318.

Yelle, Louis E., 'Adding Life Cycles to Learning Curves,' *Long Range Planning*, Vol. 16, No. 6, Dec. 1983, pp. 82–87.

<table>
<tr><td>

Sample research proposal

</td><td>

</td></tr>
</table>

This is an abridged sample of a successful proposal funded by the US National Science Foundation. An uncopyrighted public document, the layout of this sample can be adopted as a guideline for developing other research proposals. US National Science Foundation (NSF) Grant Number: DMI-9322525.

State-space and expert system hybrid model for performance measurement in design integration

Adedeji B. Badiru and Vassilios Theodoracatos
University of Oklahoma, Norman, OK, USA

RESEARCH ABSTRACT

The proposed research addresses the development of a performance measurement methodology for design integration in manufacturing systems. The methodology uses a hybrid model of state-space representation and expert system. The design process is based on the Pahl and Beitz systematic approach to design. The Triple C model of project management is used to facilitate the communication, cooperation, and coordination required for design integration. Many conceptual integration approaches are available in the literature, but most of these existing approaches lack a quantitative performance measure to drive integration efforts. The proposed performance measure is unique and the first quantifiable approach to evaluating design integration. The proposed research will facilitate the implementation of concurrent engineering concept in design and manufacturing.

PROJECT DESCRIPTION

The traditional design environment involves segregated cubicles of designers who are confined to their own worlds of design ideas. This approach to design worked in the past because consumers were less sophisticated and the market was defined more by whatever design was available to the consumers. In the present day of globalized markets, designs must be more responsive to the changing environment of the market. More of the design and manufacturing resources available within an organization must be utilized in an integrated fashion in order to create competitive designs. A competitive design will need to be not only effective, but also timely. Timeliness of design requires communication, cooperation, and coordination of design and manufacturing functions. Design information must be shared in a timely, accurate, efficient, and cost-effective manner.

The proposed research addresses the development of a performance measurement methodology for design process integration in manufacturing systems. The methodology uses a hybrid model of state-space representation and expert system. The design process is based on the Pahl and Beitz systematic approach to design. The Triple C model of project management is used to facilitate the communication, cooperation, and coordination required for design integration. Many conceptual integration approaches are available in the literature, but most of these existing approaches lack a quantitative performance measure to drive integration efforts.

The proposed research focuses on the design process rather than just the physical design itself. For the purpose of the research, we define the design process as presented below:

> The design process is the collection of all the activities and functions that support a design effort. These include the qualitative and quantitative properties of the physical design.

Literature review

Several authors have addressed the need for integration in the design environment (Preiss *et al.*, 1991; Compton, 1988; Weston *et al.*, 1991; Rzevski, 1991; Pugh, 1991; Chapman *et al.*, 1992). Several industrial enterprises and research centers are presently investigating integration methodologies. However, no effective or tangible results are in sight because there is no existing quantifiable measure of the state and progress of the design process. The proposed performance measure is unique and the first quantifiable approach to evaluating design integration.

Hundal (1991) presents a software approach to the conceptual design of technical systems. His approach involves a computer program that uses databases of requirements, functions and solutions, and evaluation techniques to generate conceptual designs. Welch and Dixon (1991) present a method for generating solutions to conceptual design problems by automated reasoning about the behavior of in-progress designs. Krishnan *et al.* (1991) present a design methodology that models cooperative design process and sequential decision strategies as a multicriteria, multilevel decision problem in the design environment. But these approaches and other similar approaches in the literature (Akagi and Fujita, 1990; Chapman *et al.*, 1992; Gebala and Eppinger, 1991; Krishnan *et al.*, 1991; Pham and De Sam Lazaro, 1990; Pugh, 1991; Sriram *et al.*, 1989; Trappey *et al.*, 1992; Weston *et al.*, 1991) are deficient in that they do not address the human elements of design process and do not offer any quantitative measure of the progress made in the integration effort.

The general characteristics of the Pahl and Beitz (1988) design approach match the desired characteristics needed to implement design integration. Trappey *et al.* (1992) present an integrated system shell, named MetaDesigner, which addresses computer-integrated product design and manufacturing planning. MetaDesigner provides a tool for the product and process designers to better relate the conceptual level of satisfying consumer needs (product requirement) and the detail level of defining design specification (physical solution).

Randhawa *et al.* (1992) present a computer-based system that integrates product design specifications with material and process databases. Using a simulation-based module, the system allows product design specifications to be evaluated in terms of economic and technical criteria. This approach permits the investigation of alternatives to design better products by integrating and exploring trade-offs between design and manufacturing. These existing approaches have the same shortcoming discussed earlier. The qualitative and quantitative techniques upon which the research will draw are available from the investigators' previous works (Badiru 1988a, 1988b, 1992; Badiru and Theodoracatos 1994).

The proposed research will facilitate the implementation of concurrent engineering concept in design and manufacturing. The state-space approach will facilitate better control of the design process at the very fundamental level. The conceptual basis for the proposed research has been documented in the literature by the principal investigators (Badiru and Theodoracatos 1994). This proposal focuses on one crucial aspect of design integration – quantitative performance measurement.

Problem statement

To achieve effective and timely design process, we propose a hybrid model of state-space representation and expert system to quantify the state and progress of the design process. The research involves an approach that models the state-to-state transformation of design process within a concurrent engineering framework. We propose for the first time, a quantifiable and measurable approach that is usable within the

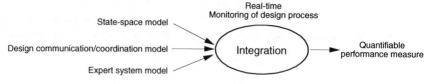

Figure 1 Design management integration model.

software, hardware, and human constraints of the design process. The integrative framework for the proposed model is illustrated in Figure 1. A state-space representation model will be the primary vehicle for linking the various components in the design process.

RESEARCH METHODOLOGY

Integrative design process

Designing is the intellectual attempt to meet certain demands in the best possible way. It is an engineering activity that impinges on nearly every sphere of human life, relies on the discoveries and laws of science, and creates the conditions for applying these laws to the manufacture of useful products. In psychological respects, designing is a creative activity that calls for a sound grounding in mathematics, physics, chemistry, mechanics, thermodynamics, hydrodynamics, electrical engineering, production engineering, materials technology and design theory, together with practical knowledge and experience in specialist fields.

In systematic respects, designing is the optimization of objectives within partly conflicting constraints. Requirements change with time, so that a particular solution can only be optimized in a particular set of circumstances. In organizational respects, designing plays an essential part in the manufacture and processing of raw materials and products. It calls for close collaboration with workers in many other spheres. Thus, to collect all the information it needs, the designer must establish close links with salesmen, buyers, cost accountants, estimators, planners, production engineers, materials specialists, researchers, test engineers and standards engineers.

An essential part of our problem-solving method involves step-by-step analysis and synthesis. In this method, we proceed from the qualitative to the quantitative, each new step being more concrete than the last. Conversion of information at each step provides data not only for the next step, but also sheds new light on the previous step.

Computer-aided design (CAD) and computer-aided manufacturing (CAM) are subsets of the design and manufacturing sub-processes of the

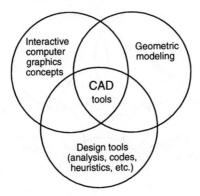

Figure 2 Interaction of CAD tools.

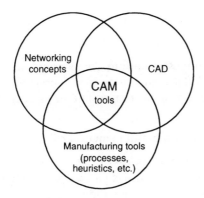

Figure 3 CAD/CAM tools interface.

general design process. CAD tools are the intersection of three sets: geometric modeling, computer graphics, and the design tools (Zeid 1991). Figure 2 illustrates the interaction of the tools. Design tools include analysis codes, such as spreadsheets, equations solvers, parametric/ variational, stress and strain, kinematic and dynamic, and fluid and thermal, heuristic procedures (Gebala and Eppinger 1991), such as expert systems, and design practices (Cutts 1988). CAM tools can be defined as the intersection of three sets: CAD tools, networking concepts, and the manufacturing tools as shown in Figure 3.

Manufacturing tools include process planning, NC programs, inspection, assembly and packaging. Process planning is the backbone of the manufacturing process since it determines the most efficient sequence to produce the product. The outcome of process planning is the production path, tools procurement, material order, and machine programming. The

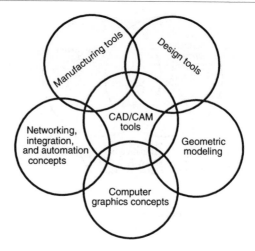

Figure 4 Expansion of CAD/CAM tools.

design process encompasses various components that must be integrated as shown in Figure 4. This integration is not possible without a quantifiable performance measure to indicate the state of each component.

Design communication, cooperation and coordination

Design management is the process of managing, allocating, and timing resources to achieve a desired design goal in an efficient and expedient manner. The objectives that constitute the goal are in terms of time, costs, and performance. Communication, cooperation, and coordination using the Triple C model of project management (Badiru 1988b) are essential in the design environment. Figure 5 shows the Triple C model applied to design control. Figure 6 shows design steps and identifies the focus of the proposed research.

Since people will be the facilitators of the design process, our project management approach will explicitly account for the human elements by using Badiru's Triple C model of project management (Badiru 1988b). The Triple C model is an effective project management approach that states that project management can be enhanced by implementing it within the integrated functions of Communication, Cooperation, and Coordination. The model facilitates a systematic approach to project planning, organizing, scheduling, and control. Figure 6 shows the essential steps for design integration.

All of the above requirements for design integration will be possible only if a reliable and quantifiable measure of the state of design is available to facilitate communication. Such a measure is provided by state-space representation.

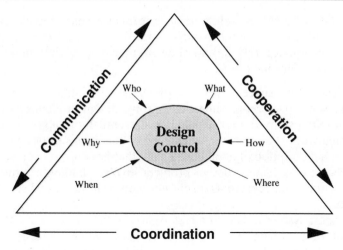

Figure 5 Triple C model for design control.

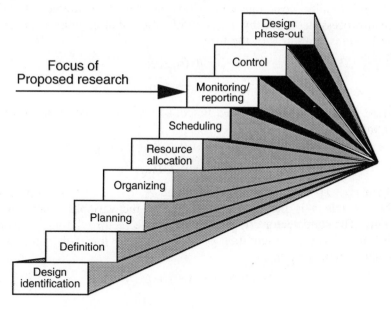

Figure 6 Design steps.

State-space representation

A state is a set of conditions that describe the design process at a specified point in time. A formal definition of state in the context of the proposed research is presented below:

The state of a design refers to a performance characteristic of the design which relates input to output such that knowledge of the input time function for $t \geq t_0$ and state at time $t = t_0$ determines the expected output for $t \geq t_0$.

A design state space is the set of all possible states of the design process. State-space representation can solve design problems by moving from an initial state to another state, and eventually to a goal state. The movement from state to state is achieved by the means of design actions. A goal is a description of an intended state that has not yet been achieved. The process of solving a design problem involves finding a sequence of design actions that represents a solution path from the initial state to the goal state.

An expert system linked to a state-space model can improve the management of a design process. A state-space model consists of state variables that describe the prevailing condition of the design process. The state variables are related to the design inputs by mathematical relationships. Examples of potential design state variables include product functionality, quality, cost, due date, resource, skill, and productivity level. For a process described by a system of differential equations, the state-space representation is of the form:

$$\dot{z} = f(z(t), x(t))$$

$$y(t) = g(z(t), x(t))$$

where f and g are vector-valued functions. For linear systems, the representation is:

$$\dot{z}' = Az(t) + Bx(t)$$

$$y(t) = Cz(t) + Dx(t)$$

where $z(t)$, $x(t)$, and $y(t)$ are vectors and A, B, C, and D are matrices. The variable y is the output vector while the variable x denotes the inputs. The state vector $z(t)$ is an intermediate vector relating $x(t)$ to $y(t)$. The state space representation of a discrete-time linear dynamic design system is represented as:

$$z(t + 1) = Az(t) + Bx(t)$$

$$y(t) = Cz(t) + Dx(t)$$

In generic terms, a design is transformed from one state to another by a driving function that produces a transitional equation given by:

$$S_s = f(x | S_p) + \varepsilon$$

where S_s is the subsequent state, x is the state variable, S_p is the preceding state, and ε is the error component. The function f is composed of a given action (or a set of actions) applied to the design process. Each inter-mediate state may represent a significant milestone in the process. Thus, a descriptive state-space model facilitates an analysis of what actions to apply in order to achieve the next desired design state.

State transformation

Design objectives are achieved by state-to-state transformation of successive design phases. Figure 7 shows the transformation of a design element from one state to another through the application of action. This simple representation can be expanded to cover several components within the design framework. The hierarchical linking of design elements provides an expanded transformation structure as shown in Figure 8.

The design state can be expanded in accordance with implicit design requirements. These requirements might include grouping of design elements, precedence linking (both technical and procedural), required communication links, and reporting requirements. The actions to be taken at each state depend on the prevailing design conditions. The natures of the subsequent alternate states depend on what actions are

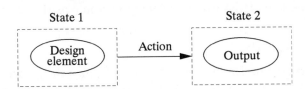

Figure 7 Transformation of design element.

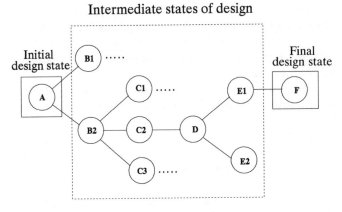

Figure 8 Design state paths.

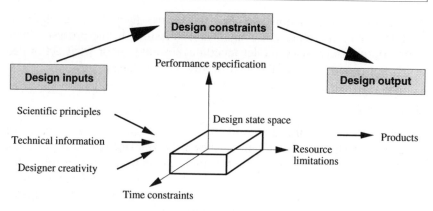

Figure 9 Factors in the design environment.

implemented. Sometimes there are multiple paths that can lead to the desired end result. At other times, there exists only one unique path to the desired objective. In conventional practice, the characteristics of the future states can only be recognized after the fact, thus making it impossible to develop adaptive plans. In the proposed research, adaptive plans can be achieved because the events occurring within and outside the design state boundaries can be taken into account as shown in Figure 9.

Real-time monitoring

If we describe a design by P state variables s_i, then the composite state of the design at any given time can be represented by a vector S containing P elements:

$$S = \{s_1, s_2, \ldots, s_P\}$$

The components of the state vector could represent either quantitative or qualitative variables (e.g. cost, energy, color, time). We can visualize every state vector as a point in the M-dimensional state space. The representation is unique since every state vector corresponds to one and only one point in the state space.

Suppose we have a set of actions (transformation agents) that we can apply to a design so as to change it from one state to another within the state space. The transformation will change a state vector into another state vector. A transformation may be a change in raw material or a change in design approach. We let T_k be the kth type of transformation. If T_k is applied to the design when it is in state S, the new state vector will be $T_k(S)$, which is another point in the state space. The number of transformations (or actions) available for a design may be finite or countably infinite. We can construct trajectories that describe the potential states of a design as we apply successive transformations. Each transformation may be repeated as many times as needed. We let T_i indicate the ith transformation in the sequence of transformations applied to the design. Given an initial state S_0, the sequence of state vectors is represented by the following:

$$S_1 = T_1(S_0)$$
$$S_2 = T_2(S_1)$$
$$S_3 = T_3(S_2)$$
$$. . . .$$
$$S_n = T_n(S_{n-1})$$

Practical example of the transformation sequence

The final state S_n depends on the initial state S and the effects of the design actions applied. We can define the following transformation sequence for a practical design process. This is a new and unique design transformation approach developed by the investigators (Badiru and Theodoracatos 1994).

S_0: | (INPUTS) Market requirements

T_1:	planning	---->> $S_1 = T_1(S_0)$:	product specification
T_2:	defining	---->> $S_2 = T_2(S_1)$:	problem statement
T_3:	formulating	---->> $S_3 = T_3(S_2)$:	overall function
T_4:	synthesizing	---->> $S_4 = T_4(S_3)$:	sub-function structure
T_5:	abstracting	---->> $S_5 = T_5(S_4)$:	basic operational structure
T_6:	varying effects	---->> $S_6 = T_6(S_5)$:	effect variants
T_7:	varying effectors	---->> $S_7 = T_7(S_6)$:	effector variants
T_8:	representing principles	---->> $S_8 = T_8(S_7)$:	solution principles
T_9:	combining	---->> $S_9 = T_9(S_8)$:	assembly variants
T_{10}:	combining	---->> $S_{10} = T_{10}(S_9)$:	system variants
T_{11}:	varying forms	---->> $S_{11} = T_{11}(S_{10})$:	varying forms
T_{12}:	laying out (arranging)	---->> $S_{12} = T_{12}(S_{11})$:	qualitative layout
T_{13}:	dimensioning	---->> $S_{13} = T_{13}(S_{12})$:	scale layout
T_{14}:	analyzing	---->> $S_{14} = T_{14}(S_{13})$:	preliminary layout
T_{15}:	elaborating	---->> $S_{15} = T_{15}(S_{14})$:	final layout
T_{16}:	detailing	---->> $S_{16} = T_{16}(S_{15})$:	detail drawings
T_{17}:	production preparation	---->> $S_{17} = T_{17}(S_{16})$:	production documents
T_{18}:	producing	---->> $S_{18} = T_{18}(S_{17})$:	product
T_{19}:	marketing	---->> $S_{19} = T_{19}(S_{18})$:	Delivery to market (OUTPUTS)

For example, $S_9 = T_9(S_8)$ indicates that by 'combining' 'solution principles,' we get 'assembly variants' where:

S_8 = solution principles (inputs)
T_9 = combining (transformation action)
S_9 = assembly variants (outputs)

State performance measurement

Ability to measure performance is essential for the success of the proposed model. A measure of design performance can be obtained at each state of the transformation trajectories. We propose a gain function $g^k(S)$ associated with the kth design transformation. The gain specifies the magnitude of enhancement (e.g. time, quality, cost savings, revenue, equipment utilization) to be achieved by applying a given design action. The difference between a gain and a performance specification is used as a criterion for determining design control actions. The performance deviation is defined as:

$$\delta = g^k(S) - p$$

where p is the performance specification. Given the number of transformations available and the current state vector, we can formulate a design policy, P, to represent the rule for determining the next action to be taken. The total design gain is then denoted as:

$$g(S|n, P) = g_1(S) + g_2(S_1) + \ldots + g_n(S_{n-1})$$

where n is the number of transformations applied and $g_i(.)$ is the ith gain in the sequence of transformations. We can consider a design environment where the starting state vector and the possible actions (transformations) are specified. We have to decide what transformations to use in what order so as to maximize the total design gain. That is, we must develop the best design process. If we let v represent a quantitative measure of the value added by a design process based on the gain function described above, then the maximum gain will be given by:

$$v(S|n) = \text{Max}\{r(S|n, P)\}$$

The maximization of the gain depends on all possible design rules that can be obtained with n transformations. Identification of the best design process will encompass both qualitative and quantitative measures in the design environment.

Non-deterministic performance

In many design situations, the results of applying transformations to the design process may not be deterministic. The new state vector, the gain generated, or both may be described by random variables in accordance with semi-Markov processes (Howard 1971). In such cases, we can define an expected total gain, $G(S_I|n)$ as the sum of the individual expected gains from all possible states starting from the initial state S_I. We let S_P be a possible new state vector generated by the probabilistic process and let $P(S_P|S_I, T^k)$ be the probability that the new state vector will be S_P if the initial state vector is S_I and the transformation T^k is applied. We can then write a recursive relation for the expected total gain:

$$G(S_I|n) = \underset{k}{\text{Max}}\left[\hat{r}^k(S) + \sum_{S_p} G(S_p|n - 1)P(S_P|S_I, T^k)\right], \quad n = 1, 2, 3, \ldots$$

where $\hat{r}^k(S)$ denotes the expected gain received by applying the kth type transformation to the design when its state is described by state vector S. In the proposed research, fuzzy modeling approach will be used for the heuristic evaluation of the gain function without resorting to esoteric mathematical analysis.

Pending research questions

We have developed the preliminary theoretical formulation of the state-space model as discussed above. This theoretical representation needs to be formulated within the context of a decision environment. Crucial areas/questions to investigate in the proposed research are:

- selection/validation of state variables;
- development of the method for generating the state equations (e.g. use of flow diagrams);
- formulation of fuzzy model representation for probabilistic states;
- delineation of data input requirements for the state-space model;
- design of the expert system implementation model.

EXPERT SYSTEM MODEL

The IF-THEN structure of knowledge representation for expert systems provides a mechanism for evaluating the changing states of design. Akagi and Fujita (1990), Rychener (1988), Pham and De Sam Lazaro (1990), and Sriram et al. (1989) have demonstrated the feasibility of using an expert system to guide engineering design. An expert system can serve as a tool that prompts the designer for the critical components of the design process (Badiru 1992; Gaarsley 1992; Mango 1992). An expert system can advise a designer through problem definition and conceptual design in the form of questions and answers for developing a task list. The expert system can also fill gaps in the conceptual design phase by producing a variant that is evaluated against existing design knowledge. Expert systems have the advantages of consistency, comprehensive evaluation of all

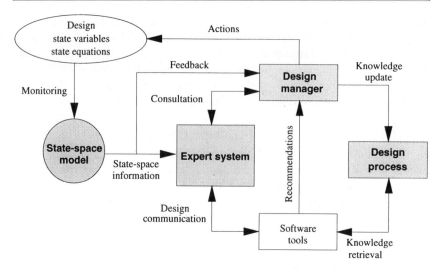

Figure 10 State-space and expert system model implementation.

available data, accessibility, infinite retention of information, and lack of bias. A pictorial representation of the proposed model is shown in Figure 10.

CLIPS (C Language Integrated Production System) will be used as the development tool for the proposed hybrid model. CLIPS is an expert system shell developed by NASA. It is a public-domain shell that has been used for many practical expert systems. It has good computational and interface capability. It can run operating system commands from within the shell, which is very useful in executing external programs. Since it is C language based, it will facilitate portability and interface with other programs.

RESEARCH AGENDA

Year one research

In year one, we will focus on the following activities:

- refinement of the state-space model;
- definition of the state variables for design state transformation;
- expansion of the list of transformation actions relating inputs to outputs;
- formulation of fuzzy model for representing probabilistic states.

Year two research

In year two, we will focus on the following:

- formal definition of measurable points in the design process;
- documentation of the design breakdown structure using the Pahl and Beitz model;
- validation of the state-space model within actual design environments.

Year three research

In year three, we will integrate the components of our model by:

- development of design rules for the expert system knowledge base;
- construction of the expert system;
- verification, testing, and validation of the hybrid model.

RESULTS OF PRELIMINARY WORK

The results of our preliminary work in the integration of project management software, expert systems, and CAD/CAM software have been very encouraging. We now have graduate students carrying out thesis research in expert system for project control, project management approach to design management, and CAD/CAM approach to design integration. An MS thesis on the integration concept was recently completed by one of the students (Ahmed 1993). Our state-space model has been successful in analytical representation of a series of design actions as previously discussed. In the proposed research, we will formalize and validate the theoretical model for integration into an expert system. The results of our preliminary work containing practical examples are presented in Badiru and Theodoracatos (1994).

MANAGEMENT PLAN

Dr Badiru will direct the project. He will supervise the design and construction of the expert system model. Dr Theodoracatos will supervise the design process component of the research. Both investigators will be involved in the mathematical formulation of the state-space model.

Dissemination of the research

The investigators will attend national conferences to disseminate the results of the proposed research. Publications in technical journals will also be prepared. They also plan to give a series of seminars on the various aspects of the research.

IMPACT OF THE PROPOSED RESEARCH

The proposed research will have a significant impact on the practical application of concurrent engineering principles, state-space modeling, and design embodiment concepts.

Effect on infrastructure of science and engineering

The proposed research will have a significant effect on infrastructure of science and engineering. It represents a new approach that combines both basic and applied research efforts to enhance the integration of design and manufacturing systems.

The research can also have an impact on the training of future engineers. New or young designers often have difficulty adapting to existing design environments that have rigid information management structures. The integrated design project management approach to be developed in this research can be used as a training tool that can help break down traditional design communication barriers.

Available facilities and equipment

The proposed investigation will be performed on the computing facilities available in the research laboratories of the investigators and the college of engineering computing network. All the necessary hardware and software are available in the CAD/CAM laboratory and the Expert Systems Laboratory, which the investigators direct. The investigators also have access to several mainframe level computer networks within the College of Engineering and the University Computing Services.

REFERENCES

Ahmed, Mohammed Faisal, *An Expert System Approach to the Domain Independent Conceptual Design Process*, MS Thesis, School of Aerospace and Mechanical Engineering, University of Oklahoma, Norman, Oklahoma, 1993.

Akagi, S. and Fujita, K., 'Building an Expert System for Engineering Design Based on the Object-Oriented Knowledge Representation Concept,' *Journal of Mechanical Design*, Vol. 112, June 1990.

Badiru, Adedeji B. and Theodoracatos, Vassilios E., 'Analytical and Integrative Expert System Model for Design Project Management,' *Journal of Design and Manufacturing*, 1994.

Badiru, Adedeji B., *Quantitative Models for Project Planning, Scheduling, and Control*, Quorum Books, Greenwood Publishing Company, Westport, CT, 1993.

Badiru, Adedeji B., *Expert Systems Applications in Engineering and Manufacturing*, Prentice-Hall, NJ, 1992.

Badiru, Adedeji B., 'State-Space Modeling for Knowledge Representation in Project Monitoring and Control,' presented at the ORSA/TIMS Fall Conference, Denver, October 1988a.

Badiru, Adedeji B., *Project Management in Manufacturing and High Technology Operations*, John Wiley, New York, 1988b.

Chapman, William L., Bahill, A. T., and Wymore, A. W., *Engineering Modeling and Design*, CRC Press Inc., Boca Raton, FL, 1992.

Compton, W. Dale, *Design and Analysis of Integrated Manufacturing Systems*, National Academy Press, Washington, DC, 1988.

Cutts, Geoff, *Structured Systems Analysis and Design Methodology*, Van Nostrand Reinhold, New York, 1988.

Gaarsley, Axel, 'Applications of Artificial Intelligence in Project Management,' *Proceedings of Annual Symposium of Project Management Institute*, Pittsburgh, PA, Sept 21–23, 1992, pp. 38–43.

Gebala, D. A. and Eppinger, S. D., 'Methods for Analyzing Design Procedures,' *Design Theory and Methodology*, DE-Vol. 31, ASME, 1991.

Howard, Ronald A., *Dynamic Probabilistic Systems: Semi-Markov and Decision Processes*, John Wiley, NY, 1971.

Hundal, M. S., 'Conceptual Design of Technical Systems: A Computer Based Approach,' *Proceedings of the 1991 NSF Design and Manufacturing Systems Conference*, Austin, TX, 9–11 January, 1991.

Krishnan, V., Eppinger, S. D., and Whitney, D. E., 'Towards a Cooperative Design Methodology: Analysis of Sequential Decision Strategies,' *Design Theory and Methodology*, DE-Vol. 31, ASME, 1991, pp. 165–172.

Mango, Ammar, 'Expert System Concepts for Project Planning,' *Proceedings of Annual Symposium of Project Management Institute*, Pittsburgh, PA, Sept 21–23, 1992, pp. 93–100.

NRC (National Research Council), Committee on Engineering Design Theory and Methodology, *Improving Engineering Design: Designing for Competitive Advantage*, National Academy Press, Washington, DC, 1991.

Pahl, G. and Beitz, W., *Engineering Design: A Systematic Approach*, edited by Ken Wallace, Springer-Verlag, NY, 1988.

Pham, D. T. and De Sam Lazaro, A., 'Autofix – An Expert CAD System for Jigs and Fixtures,' *International Journal of Machine Tool Manufacture*, Vol. 30, No. 3, 1990.

Preiss, K., Nagel, R. N., and Krenz, K. 'Design and manufacturing in an information-limited environment,' *Journal of Design and Manufacturing*, Vol. 1, No. 1, Sept 1991, pp. 17–25.

Pugh, Stuart, *Total Design: Integrated Methods for Successful Product Engineering*, Addison-Wesley, Reading, MA, 1991.

Randhawa, Sabah U., Burhanuddin, S., and Chen, Hsuen-Jen, 'An Integrated Simulation and Database System for Manufacturing Process Design Analysis,' *Journal of Design and Manufacturing*, Vol. 2, No. 1, March 1992, pp. 49–58.

Rychener, Michael D., *Expert Systems for Engineering Design*, Academic Press, Inc., New York, 1988.

Rzevski, George, 'Strategic Importance of Engineering Design,' *Journal of Design and Manufacturing*, Vol. 2, No. 1, March 1991, pp. 43–47.

Sriram, D. *et al.*, 'Knowledge Based System Applications in Engineering Design: Research at MIT,' *AI Magazine*, Vol. 10, No. 4, Fall 1989, pp. 79–96.

Trappey, Amy J. C., Liu, C. R. and Matrubhutam, S. 'An Integrated System Shell Concept for Computer-Aided Design and Planning,' *Journal of Design and Manufacturing*, Vol. 2, No. 1, March 1992, pp. 1–17.

Welch, R. V. and Dixon, J. R., 'Conceptual Design of Mechanical Systems,' *Design Theory and Methodology*, DE-Vol. 31, ASME, 1991.

Weston, R. H., Hodgson, A., Coutts, I. A. and Murgatroyd, I. S., ''Soft' Integration and its Importance in Design to Manufacture,' *Journal of Design and Manufacturing*, Vol. 1, No. 1, Sept 1991, pp. 47–56.

Zeid, Ibrahim, *CAD/CAM Theory and Practice*, McGraw-Hill, NY, 1991.

Data conversion factors

UNITS OF DATA MEASURE

Acre an area of $43\,560\,\text{ft}^2$

Agate 1/14 inch (used in printing for measuring column length)

Ampere (A): unit of electric current

Astronomical unit (AU) $93\,000\,000$ miles; the average distance of the earth from the sun (used in astronomy)

Bale a large bundle of goods (in the USA approximate weight of a bale of cotton is 500 lb) The weight may vary from country to country

Board foot 144 cubic inches (12 by 12 by 1 used for lumber)

Bolt 40 yards (used for measuring cloth)

BTU British thermal unit; amount of heat needed to increase the temperature of one pound of water by one degree Fahrenheit (252 calories)

Carat 200 mg or 3.086 troy; used for weighing precious stones (originally the weight of a seed of the carob tree in the Mediterranean region); see also Karat

Chain 66 ft (used in surveying; one mile = 80 chains)

Cubit 18 inches (derived from distance between elbow and tip of middle finger)

Decibel unit of relative loudness

Freight ton $40\,\text{ft}^3$ of merchandise (used for cargo freight)

Gross 12 dozen (144)

Hertz (Hz) unit of measurement of electromagnetic wave frequencies (measures cycles per second)

Hogshead 2 liquid barrels or 14 653 cubic inches

Horsepower the power needed to lift 33 000 lb a distance of 1 ft in 1 minute (about 1–1.5 times the power an average horse can exert); used for measuring power of mechanical engines

Karat a measure of the purity of gold; it indicates how many parts out of 24 are pure (18 karat gold is 3/4 pure gold)

Knot rate of speed of one nautical mile per hour; used for measuring speed of ships (not distance)

League approximately 3 miles

Light-year 5 880 000 000 000 miles; distance traveled by light in one year at the rate of 186 281.7 miles per second; used for measurement of interstellar space

Magnum two-quart bottle; used for measuring wine

Ohm unit of electrical resistance

Parsec approximately 3.26 light-years or 19.2 trillion miles; used for measuring interstellar distances

Pi (π) 3.14159265+; the ratio of the circumference of a circle to its diameter

Pica 1/6 inch or 12 points; used in printing for measuring column width

Pipe 2 hogsheads; used for measuring wine and other liquids

Point 0.013837 (approximately 1/72) inch or 1/12 pica; used in printing for measuring type size

Quintal 100 000 g or 220.46 lb avoirdupois

Quire 24 or 25 sheets; used for measuring paper (20 quires is one ream)

Ream 480 or 500 sheets; used for measuring paper

Roentgen (R) dosage unit of radiation exposure produced by X-rays

Score 20 units

Span 9 inches or 22.86 cm; derived from the distance between the end of the thumb and the end of the little finger when both are outstretched

Square 100 ft^2; used in building

Stone 14 lb avoirdupois in Great Britain

Therm 100 000 BTUs

Township US land measurement of almost 36 miles^2; used in surveying

Tun 252 gallons (sometimes larger); used for measuring wine and other liquids

Watt (W) unit of power

Data prefixes

Prefix	SI symbol	Multiplication factors
tera	T	$1\,000\,000\,000\,000 = 10^{12}$
giga	G	$1\,000\,000\,000 = 10^{9}$
mega	M	$1\,000\,000 = 10^{6}$
kilo	k	$1\,000 = 10^{3}$
hecto	h	$100 = 10^{2}$
deca	da	$10 = 10^{1}$
deci	d	$0.1 = 10^{-1}$
centi	c	$0.01 = 10^{-2}$
milli	m	$0.001 = 10^{-3}$
micro	μ	$0.000\,001 = 10^{-6}$
nano	n	$0.000\,000\,001 = 10^{-9}$
pico	p	$0.000\,000\,000\,001 = 10^{-12}$
femto	f	$0.000\,000\,000\,000\,001 = 10^{-15}$
atto	a	$0.000\,000\,000\,000\,000\,001 = 10^{-18}$

Abbreviations

m	meter	s	second
ft	feet	min	minute
l	liter	h	hour
g	gram	N	newton
oz	ounce	Å	angstrom
lb	pound	in	inch
W	watts	y	yard
J	joules		

Area

Multiply	By	To obtain
acres	43 560	ft^2
	4047	m^2
	4840	y^2
	0.405	hectare
cm^2	0.155	in^2
ft^2	144	in^2
	0.09290	m^2
	0.1111	y^2
in^2	645.16	mm^2
km^2	0.3861	$miles^2$
m^2	10.764	ft^2
	1.196	y^2
hectare	10 000	m^2
$miles^2$	640	acres
	2.590	km^2

Volume

Multiply	By	To obtain
acre-ft	1 233.5	m^3
cm^3	0.06102	in^3
ft^3	1 728	in^3
	7.480	gallons (US)
	0.02832	m^3
	0.03704	y^3
l	1.057	liquid quarts
	0.908	dry quart
	61.024	in^3
gallons (US)	231	in^3
	3.7854	l
	4	quarts
	0.833	British gallons
	128	US fluid oz
barrel	40	gallons
quarts (US)	0.9463	l

Mass

Multiply	By	To obtain
carat	0.200	g^3
g	0.03527	oz
kg	2.2046	lb
oz	28.350	g
lb	16	oz
	453.6	g
stone (UK)	6.35	kg
	14	lb
ton (net)	907.2	kg
	2000	lb
	0.893	gross ton
	0.907	tonne
ton (gross)	2240	lb
	1.12	net tons
	1.016	tonne
tonne (metric)	2204.623	lb
	0.984	gross ton
	1000	kg

Temperature

Conversion formulae	
Celsius to Kelvin	$K = C + 273.15$
Celsius to Fahrenheit	$F = (9/5)C + 32$
Fahrenheit to Celsius	$C = (5/9)(F - 32)$
Fahrenheit to Kelvin	$K = (5/9)(F + 459.67)$
Fahrenheit to Rankin	$R = F + 459.67$
Rankin to Kelvin	$K = (5/9)R$

Energy, heat, power

Multiply	By	To obtain
BTU	1055.9	J
	0.2520	kg- calories
W-h	3600	J
	3.409	BTU
HP (electric)	746	W
$BTU\,s^{-1}$	1055.9	W
W-s	1.00	J

Velocity

Multiply	By	To obtain
$f\,min^{-1}$	5.080	$mm\,s^{-1}$
$ft\,s^{-1}$	0.3048	$m\,s^{-1}$
$in\,s^{-1}$	0.0254	$m\,s^{-1}$
$km\,h^{-1}$	0.6214	$miles\,h^{-1}$
$m\,s^{-1}$	3.2808	$f\,s^{-1}$
	2.237	$miles\,h^{-1}$
$miles\,h^{-1}$	88.0	$f\,min^{-1}$
	0.44704	$m\,s^{-1}$
	1.6093	$km\,h^{-1}$
	0.8684	knots
knot	1.151	$miles\,h^{-1}$

Pressure

Multiply	By	To obtain
atmospheres	1.01325	bars
	33.90	f water
	29.92	in mercury
	760.0	mm mercury
bar	75.01	cm mercury
	14.50	$lb\,inch^{-2}$
$dyne\,cm^{-2}$	0.1	$N\,m^{-2}$
dyne	0.00001	N
$N\,cm^{-2}$	1.450	$lb\,inch^{-2}$
$lb\,in^{-2}$	0.06805	atmospheres
	2.036	in mercury
	27.708	in water
	68.948	millibars
	51.72	mm mercury

Length

Multiply	By	To obtain
Å	10^{-10}	m
ft	0.30480	m
	12	in
in	25.40	mm
	0.02540	m
	0.08333	ft
km	3280.8	ft
	0.6214	miles
	1094	y

Length

Multiply	By	To obtain
m	39.370	in
	3.2808	ft
	1.094	y
miles	5280	ft
	1.6093	km
	0.8694	nautical miles
mm	0.03937	in
nautical miles	6076	ft
	1.852	km
y	0.9144	m
	3	ft
	36	in

Constants

speed of light	$2.997\,925 \times 10^{10}\,\text{cm s}^{-1}$
	$983.6 \times 10^{6}\,\text{ft s}^{-1}$
	$186\,284\,\text{miles s}^{-1}$
velocity of sound	$340.3\,\text{m s}^{-1}$
	$1116\,\text{ft s}^{-1}$
gravity	$9.80665\,\text{m s}^{-2}$
(acceleration)	$32.174\,\text{ft s}^{-2}$
	$386.089\,\text{in s}^{-2}$

Bibliography

Adams, Howard G., *Making the Grade in Graduate School: Survival Strategy 101*, National Center for Graduate Education for Minorities, Notre Dame, IN, 1993.

Allen, George R., *The Graduate Students' Guide to Theses and Dissertations: A Practical Manual for Writing and Research*, Jossey-Bass Publishers, San Francisco, 1974.

Alvarez, Joseph A., *Elements of Technical Writing*, Academic Press, New York, 1986.

Badiru, Adedeji B. and Baxi, Herschel J. 'Industrial engineering education for the 21st century,' *Industrial Engineering*, Vol. 26, No. 7, July 1994, pp. 66–68.

Badiru, Adedeji B. and Pulat, P. S., *Comprehensive Project Management: Integrating Optimization Models, Management Practices, and Computers*, Prentice-Hall, Englewood Cliffs, NJ, 1995.

Barrass, Robert, *Scientists Must Write: A Guide to Better Writing for Scientists, Engineers and Students*, Chapman & Hall, London, 1978.

Bennet, John B., *Editing for Engineers*, John Wiley, New York, 1970.

Brogan, John A., *Clear Technical Writing*, McGraw-Hill, New York, 1973

Campbell, William G., Ballou, Stephen V. and Slade, Carole, *Form and Style: Theses, Reports, Term Papers*, 7th edn, Houghton Mifflin, Boston, 1986.

Coleman, P., *Technologist as Writer: An Introduction to Technical Writing*, McGraw-Hill, New York, 1971.

Davis, R. M., *Thesis Projects in Science and Engineering*, St Martin's Press, New York, 1981.

Howard, Keith, and Sharp, John A., *The Management of a Student Research Project*, Gower Publishing Company, UK, 1983.

Mauch, James E. and Birch, Jack W., *Guide to the Successful Thesis and Dissertation*, Marcel Dekker, New York, 1983.

Mitchell, John H., *Writing for Technical and Professional Journals*, John Wiley, New York, 1968.

Roth, Audrey J., *The Research Paper: Process, Form, and Content*, 5th edn, 1986.

Savory, Paul A., 'Making the move from student to teacher; Steps in the faculty search process,' *Industrial Engineering*, Vol. 26, No. 10, October 1994, pp. 60–61.

Schmidt, Steven, *Creating the Technical Report*, Prentice-Hall, Englewood Cliffs, NJ, 1983.

Sherman, Theodore A. and Johnson Simon, *Modern Technical Writing*, Prentice-Hall, Englewood Cliffs, NJ, 1983.

Smith, Robert V., *Graduate Research: A Guide for Students in the Sciences*, 2nd edn, Plenum Press, New York, 1990.

Sternberg, D., *How to Complete and Survive a Doctoral Dissertation*, St Martin's Press, New York, 1981.

Stock, M., *A Practical Guide to Graduate Research*, McGraw-Hill, New York, 1985.

Index